Petri Ahokangas · Annabeth Aagaard
Editors

The Changing World of Mobile Communications

5G, 6G and the Future of Digital Services

Editors
Petri Ahokangas
Martti Ahtisaari Institute
University of Oulu
Oulu, Finland

Annabeth Aagaard
Department of Management
Aarhus University
Aarhus, Denmark

ISBN 978-3-031-33190-9   ISBN 978-3-031-33191-6   (eBook)
https://doi.org/10.1007/978-3-031-33191-6

This Palgrave Macmillan imprint is published by the registered company Springer Nature Switzerland AG
The registered company address is: Gewerbestrasse 11, 6330 Cham, Switzerland

Foreword by Dr. Peter Stuckmann

Recent years have shown us the potential that 5G networks have to provide the connectivity basis for the digital and green recovery in the short to mid-term, and the need to build technology capacities for the following generation—6G—in the long term.

Success in 6G depends on the ability to build a resilient, secure, and high-speed 5G infrastructure, which is trusted and will support advanced 5G capabilities, on which 6G technology experiments and, later, 6G deployments can build.

The growth potential in economic activity enabled by 5G and later 6G networks and services has been estimated in the order of €3 trillion by 2030, according to McKinsey Global Institute, 2/2020.

5G networks evolution, notably 5G stand-alone, is expected to enable many industrial applications such as Connected and Automated Mobility (CAM), Industry 4.0, and advanced health care. But 6G systems will likely offer a new step change in performance, moving us from Gigabit toward Terabit capacities and sub-millisecond response times, and enabling new critical applications and an "Internet of Senses", collecting and providing the sensors data for nothing less than a digital twin of the physical world.

To explore these opportunities, the Commission launched a major initiative to promote a European vision for 6G and to develop 6G concepts, technologies and systems, the Smart Networks and Services Joint Undertaking (SNS JU). Co-led with the industry and driven by making a positive impact for our economy and society, the SNS JU addresses research, innovation, and deployment, through its two-pillar approach: it coordinates the European 5G Strategic Deployment Agenda and 5G corridors deployment projects, while fostering Europe's technology and industrial capacities in 6G, through a solid research and innovation roadmap and dedicated funding.

With its 35 newly launched 6G R&I projects worth €250 million of EU funding, and a planned investment of around the same amount in 2023 and 2024, the SNS JU is pushing for research and innovation on key technology building blocks of 6G networks. The core research is on future system architecture and control, radio and signal processing, network and service security, optical networks for converged network infrastructure, devices and components, and edge and ubiquitous computing. Smart Networks and Services projects explore how to integrate terrestrial and non-terrestrial networks, while looking at special-purpose sub-networks in very short-range communication environments. These technology capacities are expected to become the basis for future digital services toward 2030. Research is being complemented by experimental infrastructures and large-scale trials and pilots to explore and demonstrate technologies and advanced applications and services for the verticals.

Sustainability and security-by-design get special attention in 6G systems and architectures design and development, as Europe wants to lead by example the twin digital and green transition worldwide. Smart network technologies and architectures will need to drastically enhance their energy efficiency despite major traffic growth and keep electromagnetic fields (EMF) under safe limits. These design principles should form the technology base for a human-centric Next-Generation Internet (NGI) that addresses Sustainable Development Goals (SDGs) and cares about sustainability, trustworthiness, accessibility, and technology affordability, while connecting the human and digital worlds for extreme experiences.

Global standardization and spectrum harmonization need to be prepared by proactive and effective international cooperation at government and industry level. If the motto is one 6G global standard, the SNS JU needs to foster EU players' industry positions in future 6G standards and markets, building on dialogues with leading regions and possible focused joint initiatives in R&I. Also, the outcomes of EU and national 6G research projects should advise the EU-level 6G spectrum roadmap. The long-term EU spectrum strategy beyond 5G depends on a united and influential EU role in the international spectrum negotiations—notably toward the upcoming World Radio Conferences.

Moving into 6G surely presents a challenge for Europe: to maintain its technological leadership in the connectivity field, while building capacities on other fronts, such as in the edge cloud continuum, or in the chipset and components domains.

There is also a key opportunity to consolidate European technological sovereignty, in line with the conclusions of the 5G Cybersecurity toolbox. The vision is that investing and mastering 6G technologies will foster our connectivity industry, and ultimately our economy and society.

The Changing World of Mobile Communications—5G, 6G and the Future of Digital Services is timely to explore how 6G is expected to play a key role in the evolution of our economy and society toward 2030 and to contribute to a more secure and sustainable world.

Dr. Peter Stuckmann
Head of Unit—Future
Connectivity Systems
Deputy Director—
Future Networks
Interim Executive
Director—Smart Networks
and Services Joint Undertaking
European Commission

Foreword by Dr. Volker Ziegler

The commercial roll-out and evolution of 5G networks is in full swing—in the meantime, more than 240 communication service providers and thousands of enterprises have launched 5G networks. The full promise of 5G services will become reality in the next years all over the world as 5G stand-alone architecture will enable innovative offerings such as network slicing. But it does not stop there. 5G Advanced will include readiness for use cases of extended reality as well as, for instance, enhanced commercial viability and performance of IoT offerings. And 6G is already on the horizon, commercial product availability is expected in 2029/30 time frame. It will be the essential infrastructure and platform for communications in the 2030s. 6G research has successfully started with strong momentum of collaboration as illustrated by the results of worldwide relevance of the EC H2020 6G flagship project Hexa-X. Use cases of the 6G era will include immersive telepresence, twinning, collaborating robots and mixed reality co-design, creating a 6th sense and specialized trusted sub-networks. These use case families will liberate human potential and drive economic, societal, and environmental sustainability. Machine Learning and Artificial Intelligence will be pervasive enablers across the technology stack. In the 2030s, metaverse ready networks and associated new capabilities will be key. Collaborative advantage beyond research will go together with business model transformation of Network-as-a-Service and the enabling of new services by Network-as-Code. Telecom networks will become key to everything digital. Companies and organizations across every industry around the world will leverage digitalization to improve efficiency, flexibility, and productivity in a sustainable way. To meet these requirements, communication networks will be enhanced with the open flexibility and scalability of the distributed cloud. The networks of the future will sense, think, and act and thereby transform business, industry, and society. Developing Environmental, Social, and Governance (ESG) into a competitive advantage and building trustworthy platforms are strategic imperatives. Private-public partnership and regulation have been instrumental in preparing the ground for successful take-off of 5G networks. As we are now getting ready for the 6G era, the time has come to explore new paradigms from the interlinking of industrial policy, innovation, and regulation while avoiding government mandates. Availability of radio spectrum will continue to be a key prerequisite and new bands dedicated to cellular communications will help foster sustainable economic impact for the long term. Economy of scale from global standards and the avoidance of duplication of effort seem essential. Standards should be industry-lead and a fair, reasonable, and non-discriminatory patent regime will be key for value capture of practicing entities. Our purpose is to create technology that helps the world act together and connectivity continues to be the key enabler. The book *The Changing World of Mobile Communications - 5G, 6G and the Future of Digital Services* provides a comprehensive and fresh perspective on the evolution of mobile networks in the broad context of future enabling technologies and innovation, scenarios of sustainable business transformation as well as aspects of regulatory change.

Dr. Volker Ziegler
Senior Technology Advisor
Chief Architect
Nokia Strategy and Technology
Nokia

Foreword by Prof. Christopher L. Tucci

You are reading a book that is going to change the way you think about telecommunications. Telecoms has been fighting a long, slow slide from the days in which national telecoms monopolies made massive infrastructure investments and recouped them by charging high prices to businesses and consumers who had few alternatives. As these fixed-cost recovery issues have become more and more salient in recent decades with increasing deregulation and competition, there has been a shift toward mobile services as a way of adding additional revenue streams, which has helped, but recent trends indicate that telecoms companies will probably need to embrace completely new and unfamiliar business models within the next ten years as 5G continues its diffusion and 6G comes online. Both competition in telecoms services and complementor business models (such as over-the-top services) will put increasing pressure on telecoms operators.

Indeed, the rapid increase in digitalization has had an outsize impact on the way we work and live, as well as the way we communicate as individuals and as organizations. With the global pandemic, new work practices have emerged that require the ability of large numbers of people to work remotely and effectively—individually and in groups—while communicating using high-bandwidth video and other Internet applications. All of this has put increasing demands on how technologies, businesses, and services are developed in the information and communications technology (ICT) space, while at the same time pressuring ICT business models and business ecosystems. Many of the services that were previously provided by mobile network operators (MNOs) now face competition from the public Internet, challenging the business models of domestic connectivity providers. This evolution, intertwined with technology advances and regulatory bodies, builds pressure on regulators and policymakers, who must consider their responsibilities in navigating new market dynamics for the greater good of society, and who need to balance innovation promotion with protecting the public in the shorter term.

Amidst all of this complexity and co-evolution of markets, technologies, and regulations, one might sensibly ask what future world of ICT technologies, policies, businesses, and business ecosystems are on the horizon? With this edited volume, *The Changing World of Mobile Communications - 5G, 6G and the Future of Digital Services*, the editors and authors, who are among the world's leading ICT, mobile communications, and business model specialists, make a bold and critical attempt to answer this question by combining the three perspectives of business models, ICT technologies, and technology policy in an engaging and logical fashion.

In developing this narrative, the book provides cutting-edge knowledge, overviews, and unique insights into the central ICT developments that have and will have immense impact on the business, technological, and regulatory perspectives of mobile communications. In addition, the book brings forth future avenues for ICT research and mobile communications developments to guide both industry leaders, managers, researchers, and policymakers. It is difficult to predict exactly when the traditional telecoms business model will reach a breaking point, but this book will help you make an informed judgment about how soon that day will come, and what some of the options are in the new telecommunications ecosystem space.

Prof. Christopher L. Tucci
Professor of Digital Strategy
and Innovation
Imperial College
London, UK

Contents

Part I   Mapping the Mobile Communications Context

1   **Introduction to the Book**   3
    Petri Ahokangas and Annabeth Aagaard

2   **The Evolution of Mobile Communications**   13
    Seppo Yrjölä, Marja Matinmikko-Blue, and Petri Ahokangas

3   **Future Scenarios and Anticipated Impacts of 6G**   45
    Seppo Yrjölä, Petri Ahokangas, and Marja Matinmikko-Blue

4   **Sustainability Transition and 6G Mobile
    Communications**   93
    Marja Matinmikko-Blue and Ahmad Arslan

**Part II   Value Creation and Capture in Future Mobile
            Communications**

5   **Value Creation and Services in Mobile Communications**   113
    *Annabeth Aagaard, Petri Ahokangas, Marika Iivari,
    Irina Atkova, Seppo Yrjölä, and Marja Matinmikko-Blue*

6   Business Models in 5G/6G Mobile Communications   137
Petri Ahokangas, Annabeth Aagaard, Irina Atkova,
Seppo Yrjölä, and Marja Matinmikko-Blue

7   Benefiting from Innovation in Future 6G   167
Pia Hurmelinna-Laukkanen and Seppo Yrjölä

Part III   Regulatory and National Considerations

8   Local 5G/6G Network Business in Europe:
Regulatory Analysis and Legitimacy Considerations   185
Oxana Gisca, Marja Matinmikko-Blue, Petri Ahokangas,
Seppo Yrjölä, and Jillian Gordon

9   Toward Anticipatory Regulation and Beyond   221
Georg Serentschy, Paul Timmers,
and Marja Matinmikko-Blue

10   Sovereignty and 6G   253
Paul Timmers and Georg Serentschy

Part IV   Implications for the Future

11   A View to Beyond 6G   285
Seppo Yrjölä, Marja Matinmikko-Blue, and Petri Ahokangas

12   Opportunities and Implications Related to Future
Mobile Communications   307
Petri Ahokangas, Annabeth Aagaard, Seppo Yrjölä,
Marja Matinmikko-Blue, Paul Timmers,
Georg Serentschy, Jillian Gordon, Irina Atkova,
Pia Hurmelinna-Laukkanen, Ahmad Arslan,
Marika Iivari, and Oxana Gisca

Index   323

Notes on Contributors

Prof. Annabeth Aagaard earned her Ph.D. in Pharmaceutical Front-end Innovation (2009) and thereafter continued her interest and research in the innovation and business development area. She is Professor of Digitalization at the Department of Management, Aarhus University, and the Founding Director of the Interdisciplinary Centre for Digital Business Development. Her research focuses on digital and sustainable business development, open innovation & innovation management, ecosystems, and governance. She has authored and co-authored fifteen textbooks and more than 200 public and scientific papers in journals such as the *Journal of Product Innovation Management, Industrial Marketing Management*, the *International Journal of Innovation Management*, etc. She is also heavily involved in research projects in the areas of digital transformation and Next Generation Internet sponsored by EU Horizon 2020 and industrial foundations, and in addition acts as a speaker and strategic advisor to industry and Scandinavian top100 companies on digital and sustainable topics.

Prof. Petri Ahokangas received his D.Sc. degree (1998) from the University of Vaasa, Finland. Currently, he is the professor of future digital business and director of the Martti Ahtisaari Institute, Oulu Business School, University of Oulu. Prior to his academic career, he worked in the telecoms/software industry. His research is in the intersection of entrepreneurship, strategic management, international business, futures research, and action research in various fields of high technology. Specifically, he is interested in business models, strategies, ecosystems, and internationalization within digital, mobile (5G/6G), smart energy, and smart city domains. He has close to 300 scientific publications and has published in high-ranking journals such as the *IEEE Communications Magazine, IEEE Wireless communications magazine, Technological Forecasting and Social Change*, and the *Journal of Business Studies*. He actively participates in and leads international and national projects funded by EU Horizon H2020 program or national projects funded by Business Finland as primary investigator and work package leader.

Prof. Ahmad Arslan currently works as a professor at the Department of Marketing, Management and International Business, Oulu Business School, University of Oulu, Finland. He also holds the position of Honorary Chair in Business Management at the University of Aberdeen, Scotland, UK. His earlier research has been published in prestigious academic journals like the *British Journal of Management, Human Resource Management* (US), *IEEE Transactions on Engineering Management*, the *International Business Review*, the *International Marketing Review*, the *Journal of Business Research*, the *Journal of International Management*, the *Journal of Knowledge Management, Production Planning & Control, Technological Forecasting and Social Change*, and the *Scandinavian Journal of Management*, among others. He has also contributed book chapters to several edited handbooks addressing various management topics. Finally, he holds several editorial board memberships. Lastly, he is currently an Associate Editor of *International Journal of Entrepreneurship and Small Business* (Inderscience).

Assist prof. Irina Atkova received her D.Sc. degree (2018) from Oulu Business School, University of Oulu. Her dissertation explores how entrepreneurs create business models to capture opportunities. The dissertation won the 2018 Emerald/EFMD Outstanding

Doctoral Research Award in the Management and Governance category. Currently, she is a postdoctoral researcher at the Martti Ahtisaari Institute, Oulu Business School, University of Oulu, in the 6G Flagship funded by the Academy of Finland. She has published in journals such as *Entrepreneurship Theory and Practice, Strategic Entrepreneurship Journal, Global Strategy Journal,* and the Journal of Business Models. Her research interests revolve around the business model phenomenon in various applications and contexts, including telecommunications and the startup context.

M.Sc. Oxana Gisca is a Marie Skłodowska-Curie early-stage researcher within the "Legitimation of Newness and Theory Building" EU project at the Martti Ahtisaari Institute, Oulu Business School, University of Oulu. Oxana received her M.Sc. in Law at the University of the European Studies of Moldova. Prior to her academic career; she worked in a specialized agency under the Government of the Republic of Moldova and has had leading roles in international committees under the Council of Europe and the Egmont Group. Currently, she is exploring the emergence of the legitimacy of new business models and ecosystems within the 6G technology context.

Prof. Jillian Gordon is professor of Entrepreneurship at the Adam Smith Business School, University of Glasgow. Her current research examines emerging entrepreneurial ventures, the Internet of Things, digital entrepreneurship trends, and wealth recycling. She is particularly interested in how the practice and logics of entrepreneurship translate across social contexts. Her research has been published in leading international journals, including the *Harvard Business History Review, Business History,* and FT 50 ranked journals, including the *Journal of Business Ethics* and *Human Relations.* During her career, she has also secured several competitive research funding awards from major funders, including the ESRC, EPSRC, Horizon 2020, and the EU EIT Scheme.

Prof. Pia Hurmelinna-Laukkanen is a professor and director of the Department of Marketing, Management and International Business at the Oulu Business School, University of Oulu, and an Adjunct Professor at LUT University. She has published about 90 refereed articles in journals such as the *Journal of Management Studies, Research Policy,* the *Journal of Product Innovation Management,* the *California Management Review, Industrial and Corporate Change, Industrial Marketing Management,* the *International Business Review, R&D Management,* and *Technovation.* Most of her research has involved innovation management and appropriability issues, including an examination of different knowledge protection and value capturing mechanisms and their strategic uses. Her research covers varying contexts like internationalization and interorganizational collaboration in ICT and the healthcare sectors.

Dr. Marika Iivari is an Adjunct Professor in the field of Digitalization and Business Analytics at the Martti Ahtisaari Institute, Oulu Business School, University of Oulu. She gained her D.Sc. (Econ. & Bus. Admin.) in the field of business models, open innovation and ecosystems in 2016, and has worked in several national and international R&D&I projects in various technological contexts. She has been an advisor in the Urban Agenda for EU's Digital Transition Partnership. In addition, she has industry experience, having worked in management roles in the software business. She is also associated with Imperial College London's Centre for Digital Transformation. Currently, her research interests, in addition to innovation, business modeling and design, expand to data-driven 106 decision making and the data economy.

Dr. Marja Matinmikko-Blue received her D.Sc. degree (2012) in telecommunications engineering and Ph.D. degree (2018) in industrial engineering and management from the University of Oulu, Finland. Currently, she is Research Director of the Infotech Oulu Focus Institute and Director of Sustainability and Regulation at the 6G Flagship program at the University of Oulu, where she also holds an adjunct professor position on spectrum management. She conducts multidisciplinary research on technical, business, and regulatory aspects of mobile communications systems in close collaboration between industry, academia, and regulators. She has coordinated four national project consortia that have successfully demonstrated the world's first licensed shared access spectrum sharing trials and introduced a new local 5G operator concept that has become a reality. She has published 190+ scientific publications and prepared 150+ contributions to regulatory bodies on spectrum management at national, European, and international levels.

Dr. Georg Serentschy with a professional span of more than 40 years, began his career in the field of nucle physics, after which he devoted himself to industrial research and development in various industrial high-tech areas such as software development, solar energy, aerospace, and telecommunications. After his industrial career, he joined Arthur D. Little, a strategy consulting firm. His next step was to head the Regulatory Authority for Telecommunications in Austria (RTR-GmbH) for more than a decade. The highlight of his regulatory career was the chairmanship and vice-chairmanship of BEREC (Body of European Regulators for Electronic Communications). In 2014, he founded his own consulting boutique, which focuses on advising the C-suite and top experts in the digital sector (platforms, telecom, media, and technology) on strategy, artificial intelligence, quantum technologies, regulation & competition, spectrum policy, cybersecurity policy, and innovation. One of the focus areas of his consulting work is the strategic positioning of companies in a specific regulatory environment. In parallel, Georg serves as Senior Advisor for the Communication Practice and the Public Policy Practice of the global law firm Squire Patton Boggs.

Prof. Dr. Paul Timmers is research associate at the University of Oxford, Oxford Internet Institute, professor at KU Leuven and European University Cyprus and the University of Rijeka (visiting), senior advisor EPC Brussels, President of the Supervisory Board Estonian eGovernance Academy, member of the EU Cyber Direct Advisory Board, research fellow of CERRE, and CEO of iivii. Previously, he was Director at the European Commission/DG CONNECT where has held responsibility for legislation and funding programmes for cybersecurity, e-ID, digital privacy, digital health, smart cities, and e-government. He was also a cabinet member of European Commissioner Liikanen. He worked as manager of a software department in a large ICT company and co-founded an ICT start-up. He holds a physics PhD from Radboud University (Nijmegen, NL), MBA from Warwick University (UK), EU fellowship at UNC Chapel Hill (US), and a cybersecurity qualification from Harvard. His main interests are digital policy, geopolitics, and

Europe. He frequently publishes and speaks on digital developments, technology and sovereignty, cybersecurity, industrial policy, and sectoral policies such as digital health, and is regularly advising governments and thinktanks.

Prof. Seppo Yrjölä holds a D.Sc. degree in Telecommunications Engineering from the University of Oulu and is Professor at the faculty of Information Technology and Electrical Engineering, University of Oulu in the field of techno-economics. He is Principal Engineer at Nokia Enterprise and has been building radios for 34 years in research, development, innovation, and business development. Previously, he was head of wireless technology for the Networks division at Nokia. He has been awarded over €10 million in competitive research funding as the lead investigator, authored more than 140+ scientific publications, and holds several patents in the radio domain. Prof. Yrjölä conducts multidisciplinary research, combining technology, business, and regulatory aspects. His current mission is digitalizing the 70% of GDP that has yet to be digitalized in order to drive massive productivity growth and new business. With roots in engineering and economics, he explores how and why platform-based ecosystemic business models can emerge in the future wireless system context in a sustainable and human-centric way.

Abbreviations

3 Cs	Capacities, Capabilities, Control
3D	Three Dimensional
3GPP	3rd Generation Partnership Project
3GPP2	3rd Generation Partnership Project Two
4C	Connectivity, Content, Context, Commerce
5GA	The Fifth Generation-Advanced Mobile Communication
5GC	The Fifth Generation Core Network
5C	Connectivity, Cloud, Content, Context, Commerce
AaaS	Application-as-a-Service
aaS	as a Service
AI	Artificial Intelligence
AIA	Artificial Intelligence Act
API	Application Programming Interface
AR	Augmented Reality
ATIS	US Alliance for Telecommunications Industry Solutions
B2B	Business to Business
B2C	Business to Consumer
B5GPC	Japan's Beyond 5G Promotion Consortium
BATX	Baidu, Alibaba, Tencent, Xiaomi
BEREC	Body of European Regulators for Electronic Communications

BM	Business Model
BMI	Business Model Innovation
BRI	Chinese Belt and Road Initiative
BTS	Basic Telecommunications Services
CAPEX	Capital Expenditure
CDMA2000	Code Division Multiple Access 2000
CECC	European Electronic Communications Code
CEPT	European Conference of Postal and Telecommunications Administrations
CLA	Causal Layered Analysis
CMA	UK's Competition and Markets Authority
CO2	Carbon Dioxide
COVID-19	Coronavirus Disease
CS	Cybersecurity
CSA	Cybersecurity Act
CSL	China's Cybersecurity Law
CT	Core Network & Terminals
CV	Curriculum Vitae
DA	Data Act
DaaS	Data as a Service
DAO	Decentralized Autonomous Organization
DC	EU's Digital Compass
DDI	Digital Dependency Index
DevOps	Development and Operations
DG	Directorate General
DGA	Data Governance Act
DMA	Digital Market Act
DNA	Deoxyribonucleic Acid
DRCF	UK's Digital Regulation Cooperation Forum
DS	Digital Strategy
DSA	Digital Service Act
DSM	Digital Single Market
e2e	end to end
EC	European Commission
EC DSM	EU's Digital Single Market Strategy
EC ePR	e-Privacy Regulation
EC ODD	EU's Open Data Directive
EC RPE	EU's recovery Plan for Europe
EC SEDF	EU's shaping Europe's Digital Future

ECJ	European Court of Justice, Luxembourg
EDE	European Digital Economy
EECC	European Electronic Communications Code
eID	electronic Identification
EMA	Singapore's Energy Market Authority
eMBB	enhanced Mobile Broadband
EMF	Electromagnetic Field
ENISA	European Cybersecurity Agency
ESG-D	Environmental, Social, Governance, Democracy
ETNO	European Telecommunications Network Operators' Association
ETSI	European Telecommunications Standards Institute
EU	European Union
FAANG	Facebook, Amazon, Apple, Netflix, Alphabet
FP	Framework Program
FRAND	Fair, Reasonable, and Non-Discriminatory
GAFA	Google, Apple, Facebook, Amazon
GDP	Gross Domestic Product
GDPR	EU's General Data Protection Regulation
GHG	Greenhouse Gas
GPT	General-Purpose Technology
GSM	Global System for Mobile Communications
GSMA	Global System for Mobile Communications Association
HMI	Human-Machine Interface
HW	Hardware
I5.0	Industry 5.0
IaaS	Infrastructure as a Service
IAB	Internet Architecture Board
IaC	Infrastructure as Code
IASB	International Accounting Standards Board
ICANN	Internet Corporation for Assigned Names and Numbers
ICDT	Information, Communication, and Data Technology
ICO	UK's Information Commissioner's Office
ICT	Information and Communication Technology
IEEE	Institute of Electrical and Electronics Engineers
IETF	Internet Engineering Task Force
IIoT	Industrial Internet of Things
IMDA	Singapore's Infocomm Media Development Authority
IMT	International Mobile Telecommunications

IoT	Internet of Things
IIoT	Industrial Internet of Things
IP	Intellectual Property
IPR	Intellectual Property Right
ISP	Internet Service Provider
IT	Information Technology
ITU	UN's International Telecommunication Union
ITU-R	UN's International Telecommunication Union Radiocommunication Sector
ITU-T	UN's International Telecommunication Union Telecommunication Standardization Sector
KFTC	Korea Fair Trade Commission
KPI	Key Performance Indicator
KVI	Key Value Indicator
LEO	Low Earth Orbit
LoRaWan	Long Range Wide Area Network
LTE	Long-Term Evolution, the Fourth Generation of Mobile Communications
MAS	Money Authority of Singapore
MIIT	China's Ministry of Industry and Information Technology
ML	Machine Learning
MLP	Multi-Level Perspective
mMTC	Massive Machine Type Communication
MNO	Mobile Network Operator
MOST	China's Ministry of Science and Technology
MRFTA	Korea's Monopoly, Regulation and Fair-Trade Act
MSIT	South Korea's Ministry of Science and ICT
MVNO	Mobile Virtual Network Operator
NaaC	Network as a Code
NaaS	Network as a Service
NBIC	Nanotechnology, Biotechnology, Information Technology, and Cognitive Science
NCC	Taiwan's National Communications Commission
nG	nth Generation Of Mobile Communications (1G-6G)
NGA	NextG Alliance
NGO	Non-Governmental Organization
NICT	Japan's National Institute of Information and Communications Technology
NPN	Nonpublic Network

NRA	National Regulatory Agency
NRAs	National Regulatory Authorities
NSA	Non Stand Alone
NTN	Non Terrestrial Network
OEM	Original Equipment Manufacturer
OFCOM	UK's Office of Communications
OPEX	Operational Expenditure
OTT	Over the Top
P2P	Peer to Peer
PaaS	Platform as a Service
PESTLE	Political, Environmental, Social, Technological, Legal, and Economic
PFI	Profiting From Innovation
PNI	Public Network Integrated
PPP	Public-Private-People
PRC	People's Republic of China
R&D	Research and Development
RAN	Radio Access Network
RPE	EU's Recovery Plan for Europe
SA	Stand Alone or System Architecture
SaaS	Software as a Service
SBA	Service-Based Architecture
SDGs	UN's 17 Sustainable Development Goals
SDN	Software Defined Radio
SEDF	Shaping Europe's Digital Future
SEP	Standard Essential Patent
SLA	Service Level Agreement
SNSJU	European Smart Networks and Services Joint Undertaking
SU	Security Union
SW	Software
SWIFT	Society for Worldwide Interbank Financial Telecommunication
TBL	Triple Bottom Line
TCO	Total Cost of Ownership
TDMA	Time Division Multiple Access
THz	Terahertz
TIP	Transformative Innovation Policy
TTC	Trans-Atlantic Trade and Technology Council
UE	User Equipment
UI	User Interface

UMTS	Universal Mobile Telecommunications System
UN	United Nations
URLLC	Ultra-Reliable Low-Latency Communications
US	United States of America
VATS	Value-Added Telecommunications Services
VR	Virtual Reality
VUCA	Volatility, Uncertainty, Complexity, and Ambiguity
W3C	World Wide Web Consortium
Web3	World Wide Web Three
Wi-Fi	Wireless Fidelity
WIMAX	Worldwide Interoperability for Microwave Access
WP5D	ITU-R's Working Party 5D
WRC	ITU-R's World Radio Conference
XaaS	Everything as a Service
XR	Extended Reality

List of Figures

Fig. 2.1 Expected 3GPP standardization timeline and ITU-R process for IMT systems 17

Fig. 2.2 The transformation of technology innovation across 6G system architecture layers stems from the 3rd Generation Partnership Project (3GPP) system architecture (Adapted from Yrjölä et al., 2022) 23

Fig. 2.3 The value chain in 2G, 3G, 4G (upper part of the figure), and 5G (lower part of the figure) 28

Fig. 2.4 From engineering platforms toward service modularity and ecosystem platforms 29

Fig. 2.5 Evolution from vertical and horizontal business logics toward oblique value creation and value capture in 6G (Adapted form Yrjölä et al., 2022) 34

Fig. 3.1 Identified key trends 48

Fig. 3.2 Identified key uncertainties 55

Fig. 3.3 Summary of selected scenario themes 61

Fig. 3.4 Summary of four scenarios 62

Fig. 6.1 A strategy-technology view on mobile communications business models 151

Fig. 6.2   Three types of business model and ecosystem
            configurations in 6G   159
Fig. 7.1   Benefiting from innovation in a 6G context
            framework with a comparison to 4G and 5G (*Source*
            Hurmelinna-Laukkanen and Yang, [2022])   171
Fig. 8.1   The EU's priorities for the digital single market   195
Fig. 8.2   Regulatory legitimacy challenges and perspectives   208
Fig. 9.1   Regulatory trajectories from traditional to anticipatory
            regulation   240
Fig. 10.1   Geopolitics versus technology   263
Fig. 11.1   A legitimation view of the worldviews, myths,
            and metaphors in 6G visions   300

List of Tables

Table 5.1   Identified mobile communications services in 5G   121
Table 8.1   Managerial choices and consequences derived
             from the identified legal framework for emerging
             local 5G/6G private networks   213
Table 9.1   5th Generation Regulation (G5) countries by score,
             rank, and the ICT regulatory tracker   238
Table 10.1   Recommendation toward open, global, and full 6G   273
Table 10.2   Participation of Chinese companies in EU-funded
             projects   279
Table 11.1   The causal layered analysis of national and regional
             6G visions transformed into 6G futures toward
             beyond 6G   297

Part I

Mapping the Mobile Communications Context

1

Introduction to the Book

Petri Ahokangas and Annabeth Aagaard

I have a story to tell you. It has many beginnings, and perhaps one ending. Perhaps not. Beginnings and endings are contingent things anyway; inventions, devices. Where does any story really begin? There is always context, always an encompassingly greater epic, always something before the described events, unless…

(Iain M. Banks, The Algebraist)

P. Ahokangas (✉)
Martti Ahtisaari Institute, University of Oulu, Oulu, Finland
e-mail: petri.ahokangas@oulu.fi

A. Aagaard
Department of Management, Aarhus University, Aarhus, Denmark
e-mail: aaa@mgmt.au.dk

© The Author(s) 2024
P. Ahokangas and A. Aagaard (eds.), *The Changing World of Mobile Communications*,
https://doi.org/10.1007/978-3-031-33191-6_1

The Aim and Purpose of the Book

Mobile communications as the backbone for digitalization in modern society in many ways define how digital services are being designed, delivered, and consumed. Despite its central role for individuals, organizations, and societies in digitalization, the mobile communications context is only scantily researched outside the engineering domain. However, the idea for the book was initiated within the world's first 6G research program, the **6G Flagship** at the University of Oulu, Finland, funded by the Academy of Finland as an eight-year research endeavor from 2018 to 2026. The book aims to provide a comprehensive and multidisciplinary outlook on the present and future, focusing on the changing world of mobile communications, written by a team of authors representing relevant experience and expertise in the *business, regulation, and technology* management domains. Contrary to many conventionally edited and peer-reviewed scientific books, the content of the chapters have been peer-reviewed and coordinated to provide a coherent, holistic, and multidisciplinary forward-looking view to understand and make sense of what we call the world of mobile communications.

Mobile communications technologies are often referred to by examining what generation of technology they represent in the continuum from the first generation (1G) to the latest fifth generation (5G) technology. These generations are backward compatible, meaning that the currently used communications technologies may be based on any of the generations from 2G to 5G. These technologies should also be forward compatible with the next generation mobile communications technologies. The research on future 6G already started a few years ago, and we are expecting that it will be commercially available by 2030. However, what 6G will be, and how it will be used, remains unknown.

Today's 4G services are available practically everywhere, and the adoption of 5G networks is well underway. Compared to 4G, 5G has already brought about new business opportunities, especially in industrial domains and by enabling seamless virtual and augmented reality services. However, it (5G) also raised serious concerns about data privacy and security and the use of artificial intelligence. As the global vision for 6G will be released in 2023, we need to understand already today what 5G evolution and 6G may bring for the future service delivery, and how they will influence us at user, business, business ecosystem, and geopolitical levels. Future 5G evolution and 6G are not only about moving toward faster, better, and more secure networks providing the backbone for innovative digital services. 5G and 6G will bring about a profound digital disruption that concerns everybody: individuals as the users and the developers of the service; companies that are developing and providing these technologies; companies that are providing mobile communications services, i.e., the operators; companies that are offering their services on public or private 5G and 6G services or needing these services; the authorities and public organizations; and policymakers and regulators.

Currently, no books are available that bring together business and regulation perspectives of mobile communications with strong engineering expertise. A more holistic picture of 5G and 6G is therefore lacking in existing publications and in the present discussions of ICT. Consequently, the aim and purpose of the book it to provide the reader with a state-of-the-art, multidisciplinary, and insightful overview and vision presented by experts in the field.

With this book, we, therefore, aim to explore and provide answers to the following questions:

- What will 5G, its evolution, and 6G be about?
- How will 5G and 6G influence future digital services, businesses, and society, and what kind of impacts will they have on them?
- How should 5G and 6G be regulated in the future?
- How could we benefit from 5G and 6G innovations in the future?

Who Should Read the Book, and Why?

This book is targeted at and written for managers, practitioners, policymakers, and students who want to understand what 5G and 6G will be about, the kind of impacts they might have, and how we can benefit from them in the future. It also provides a holistic view of future mobile communications business and regulative aspects to engineers working in the mobile communications sector. The key subject areas concerned are engineering, policy, management, and business. The book can be used in higher education in engineering and management, as well as digital business. Furthermore, the book is applicable by (1) managers, who are active in mobile communications and/or apply mobile connectivity as the backbone of their digitalization and digital services, (2) regulatory bodies and policymakers, who operate in the field of mobile connectivity, or whose work has implications for related businesses and services, and (3) researchers and universities and other higher education institutions, and their students. Accordingly, the book:

- Provides a holistic and insightful view of the future of mobile connectivity as the backbone for all digitalization, given by experts in the field.
- Combines the technical and business-related perspectives of the field in exploring the unique and vast business potential, while addressing the impact on policymaking.
- Contributes valuable insights and new knowledge to all who study, develop, manage, provide, use, and regulate mobile connectivity and related businesses.
- Inspires the reader through practical and reality-based examples and industry views on future 5G/6G.

How Was the Book Developed?

With the aim to provide a more holistic view of 5G, 6G, and beyond, in leveraging more human-centric information and communications technologies (ICT), we have initiated, developed, and edited this book on *The Changing World of Mobile Communications: 5G, 6G, and the Future of Digital Services* with chapters co-authored by researchers and practitioners working at the interface between business, engineering, and policymaking. Through close dialogues and collaborations between the co-authors, we have aimed to facilitate a more holistic discussion and presentation of the content, impact, and future scenarios of telecommunications. In doing so, the book seeks to bridge these three large and vibrant research communities with a view to informing future research, as well as practitioners and policymakers, on the neglected but vital contributions that ICT ecosystems and businesses can make to sustainably creating and capturing value for society. Underpinning the book is the core question of how 5G/6G can contribute to sustainable value creation and value capture from a business, engineering, and policymaking perspective.

Structure of the Book

The structure of the book is created to answer these questions and build a stepwise learning experience for the reader in four parts. Part I maps the mobile communications context, Part II examines value creation and capture in the context, Part III focuses regulatory and national considerations, and Part IV, building collectively on the preceding chapters, discusses implications for future consideration for research, management, and policymaking.

Part I Mapping the Mobile Communications Context

Part I of the book starts with an introduction (Chapter 1). The *Introduction to the book* chapter provides an overview of the book's purpose, aim, content, and targeted contribution and target audiences and provides a short presentation to each chapter and biographies of all the authors.

Chapter 2 *Evolution of mobile communications* introduces the evolution of mobile communications. As this context is expected to become increasingly platform-based and ecosystemic, it is important to distinguish relevant perspectives to map the developments in the field. The chapter provides an outlook from the first (1G) to fifth (5G) generation of mobile communications by examining technology and standardization, relevant regulatory developments, and content, and specifically characterizes the business ecosystems toward the sixth generation of mobile communications (6G). The purpose of the chapter is to provide a contextual setting for the discussions presented in the subsequent chapters by showing the emergence and evolutionary continuum of mobile communications from 1G toward 6G.

Chapter 3 *Future scenarios and anticipated impacts of 6G* examines future scenarios of 6G at different levels of analysis, aiming to identify and assess the key change political, environmental, social, technological, and legal forces—trends and uncertainties—related to future mobile communications and proposes a set of dimensions according to which we can expect 6G to change the world. Based on the proposed dimension, the chapter presents a set of future scenarios related to mobile connectivity integrated with various services at the user (humans and machines), business (service provisioning and utilization), business ecosystem value chain (upstream and downstream), and geopolitical levels of analysis.

Chapter 4 *Sustainability transition and 6G mobile communications* highlights 6G mobile communications' link with the sustainability transition. Using both theoretical arguments and practical examples, the chapter applies the multilevel perspective of the sustainability transition to highlight specificities of the niches, sociotechnical regimes, and exogenous sociotechnical landscapes of 6G technology in relation to the sustainability transition.

Part II Value Creation and Capture in Future Mobile Communications

Chapter 5 *Value creation and services in mobile communications* explores the opportunities for value creation via the services enabled by the fifth (5G) and sixth (6G) generation of mobile communications, with a specific focus on value creation at the service, platform, and ecosystem levels of analysis. The chapter presents the 5G and 6G usage cases as starting points, highlighting the drivers of value creation and the key services enabled by the mobile communications technology generations.

Chapter 6 *Business models in 5G/6G mobile communications* examines business models in mobile communications that have remained surprisingly stable up to 4G. 5G and beyond generations will bring a fundamental change to how mobile connectivity is deployed and commercialized. This chapter explores the business opportunities, business models, and changing platformic business ecosystems of the future that extend beyond traditional company boundaries.

Chapter 7 *Benefiting from innovation in future 6G* takes a wider perspective on profiting from innovation and discusses firm, ecosystem, industry, and policy-level aspects relevant for developing 5G/6G. To date, mobile communications networks have been seen as enabling technologies whose innovations potential can be characterized by examining technology complementarity, standardization, and intellectual property issues. With 6G, especially in combining artificial intelligence, the mobile network gains features of a general-purpose technology platform with specificities regarding the appropriability of value.

Part III Regulatory and National Considerations

Chapter 8 *Local 5G/6G network business in Europe: regulatory analysis and legitimacy considerations* focuses on the EU Digital Legal Framework; it identifies and discusses relevant EU legal acts and presents the EU legal initiatives in the context of local mobile communications networks. It reviews previous research from the legitimacy challenge perspective and improves understanding of how regulation currently delimits the emerging business models of the local 5G/6G networks.

Chapter 9 *Toward anticipatory regulation and beyond* discusses the need of telecom regulators to break out of their sectoral silos and analyzes the regulatory situation in Europe, North America, and Asia. The chapters derive conclusions on how Europe's competitiveness and innovative strength can be improved with better interlinking of industrial policy, innovation, and regulation by introducing the anticipatory regulation approach.

Chapter 10 *Sovereignty and 6G* discusses how sovereignty has become a top priority for government leaders and asks what sovereignty is in the world of 6G. Or is the question rather what 6G is in a world where safeguarding sovereignty is the major theme of geopolitical collaboration, competition, and conflict? The chapter outlines the interplay of 6G technology and political-industrial governance in different scenarios for the future of sovereignty.

Part IV Implications for the Future

Chapter 11 *A view to beyond 6G* adopts a futures research approach and applies causal layered analysis to presented 6G visions, focusing on different national perspectives between China, Europe, Japan, South Korea, and the USA. The chapter identifies the assumptions behind mobile communications, analyzes the different national visions and presents, based on the multiple ideologies and epistemes of the stakeholders, transformed futures beyond 6G mobile communications. The chapter concludes with policy implications for developing global mobile communications.

Finally, summarizing the discussions in the preceding 11 chapters, Chapter 12 *Opportunities and implications related to future mobile communications* focuses on the research, managerial, and policymaking opportunities and implications related to future mobile communications.

Acknowledgements Writing and editing *"The Changing World of Mobile Communications: 5G, 6G and the Future of Digital Services"* has been one big team project of editors and co-authors working diligently together on developing the chapters, while coordinating and sharing knowledge in making state-of-the-art contributions for a holistic publication on the emergence, impact, and scenarios of 5G and 6G. Accordingly, huge thanks are due to all the book's co-authors!

This research has been supported by the Academy of Finland, 6G Flagship program under Grant 346208. This book would not have been possible without the funding, support, and the network and especially the community of the 6G Flagship program at the University of Oulu, Finland. The 6G Flagships builds on the *"6G vision for 2030: Our future society is data-driven, enabled by near-instant and unlimited wireless connectivity. Developing products, services and vertical applications for the future digitized society requires a multidisciplinary approach and a re-imagining of how we create, deliver and consume data and services.*" A special thank you Matti Latva-aho and Ari Pouttu for the support and funding provided!

In addition, this book has been partially supported by the LNETN project (Legitimation of newness and its impact on EU agenda for change) from the European Union's Horizon 2020 research and innovation program under the Marie Skłodowska-Curie grant agreement No. 860364.

However, "behind the scenes" of the book are several key people who have played vital roles in its making. We would like to thank Alec Selwyn and Arunaa Devi from Palgrave MacMillan for a very fruitful and effective collaboration on developing, editing, printing, and launching the book. Second, we want to thank Bea Longhurst for all her hard work in remaking and redesigning all the figures of the book and for achieving a professional and unified infographic look. Finally, we thank Gareth Attwood and Rupert Moreton from Acolad for their professional and effective high-quality proofreading of the chapters.

2

The Evolution of Mobile Communications

Seppo Yrjölä, Marja Matinmikko-Blue,
and Petri Ahokangas

Natural selection will not remove ignorance from future generations.

(Richard Dawkins)

S. Yrjölä (✉)
Centre for Wireless Communications, University of Oulu, Oulu, Finland
e-mail: seppo.yrjola@oulu.fi; seppo.yrjola@nokia.com

Nokia, Oulu, Finland

M. Matinmikko-Blue
Infotech Oulu Focus Institute and Centre for Wireless Communications,
University of Oulu, Oulu, Finland
e-mail: marja.matinmikko@oulu.fi

© The Author(s) 2024   **13**
P. Ahokangas and A. Aagaard (eds.), *The Changing World of Mobile Communications*,
https://doi.org/10.1007/978-3-031-33191-6_2

The Historical Development of Mobile Communications

Worldwide digitalization has been enabled by the successive mobile communications generations over the past three decades. Each generation has introduced new use cases and technical capabilities, while optimizing the use cases of the previous generation. Overall, technology can be seen to serve an enabling role in mobile communications. In this historical development, the commercialization cycle of mobile communications has followed three steps: (1) definition, (2) standardization and implementation, and (3) deployment and use. At the *definition* stage, the innovation from companies and research organizations is mediated together with national authorities in the global ITU-R (International Telecommunication Union Radiocommunication sector) to form a framework and develop usage scenarios for the radio aspects of mobile communications technology. After the definition of the requirements for the radio interface at the ITU-R, standardization bodies and firms negotiate *standardization and implementation* via standard releases that are the basis for the implementations by different technology vendors. As the technical systems and solutions needed in a technology generation have been developed, they are deployed and utilized by the mobile operators in different business implementations (Ahokangas et al., 2023). This *deployment and use* are, however, delimited by regulation as the telecommunications is a highly regulated field.

This chapter provides an overview and brief introduction to the mobile communications industry. The chapter will start with a brief description of technological development in the field in different technology generations from the first to sixth generation and discuss the role of standardization in this development. Next, the chapter will provide

P. Ahokangas
Martti Ahtisaari Institute, Oulu Business School, University of Oulu, Oulu, Finland
e-mail: petri.ahokangas@oulu.fi

a short introduction to the role of regulation in mobile communications. The chapter will conclude with a characterization of the mobile communications business. This chapter serves as a starting point for the discussions presented in the subsequent chapters.

Technological Developments from 1G to 5G

From the first generation, the mobile network system architecture has been defined by the radio access technology, access and core network routing, and the associated services related to voice, messaging, data transfer, mobility, authentication, and access control. After the first generation (1G) analog voice only service, the second generation (2G) introduced a digital mobile system with text messaging and mobile phones as a personal portable device in addition to a voice service. The third generation (3G) with mobile broadband data brought access to mobile multimedia and significantly lowered the cost of the voice service. The fourth generation (4G) expanded the multimedia service offering across digital industries built around smart phones. 4G lowered the cost of data while introducing video to consumers and machine-type communications to serve vertical industries. The ongoing deployment of the fifth generation (5G) has drastically increased the number of communicating objects (David & Berndt, 2018). For consumers interactive low-cost video and for enterprises the industrial IoT (Internet of Things) are paving the way toward human augmentation and digital-physical fusion. Up to 4G mobile communications, the connectivity business has remained surprisingly unchanged allowing the incumbent mobile network operators (MNOs) to dominate the market, although they have been seriously challenged by the content-owning, cloud-based over-the-top (OTT) Internet giants (Ahokangas et al., 2013).

For 5G, the ITU-R vision framework for international mobile telecommunications IMT-2020 and beyond presented in (ITU-R, 2020) adopted a service-centric approach to the 5G use case definition. The IMT-2020 vision identified three services classes, enhanced mobile broadband (eMBB) targeted at consumers, ultra-reliable low-latency communications (URLLC) for mission-critical services for organizations such as factories, and massive machine-type communications (mMTC) to connect IoT. The fifth generation mobile network 5G new radio (NR) solution was standardized by 3GPP in release 15 and commercially deployed in 2019 based on the non-standalone (NSA) architecture where a 5G radio access network (RAN) operates on a legacy 4G LTE core network. Innovations in a new user equipment (UE), radio access network (RAN), and 5G core (5GC) designs enable substantial improvements across the main service domains eMBB, URLLC, and mMTC. In particular, the new active antenna beam-based physical layer RAN design allows operation in higher frequencies up to mmWaves with wider bandwidths. The 5G system architecture including the 5GC became available from 2020 as standalone (SA) enabling the deployment of private enterprise and industrial 5G networks (Parkvall et al., 2017).

The 5G standard evolution in releases 16 and 17, as depicted in Fig. 2.1, expands the 5G ecosystem particularly for industrial domain via innovations such as time sensitive communication, small data transmission, and UE energy saving. 3GPP work on release 18 5G-Advanced (5GA) is due in 2024 and the first deployments are expected around 2025 (Chen et al., 2022). 5GA will provide an intelligent network platform utilizing machine learning (ML) to adapt to its environment, new classes of devices and enhance support for novel applications such as truly mobile extended reality (XR) services. Furthermore, 5GA will embed high-precision location, presence and timing technologies, and device innovations will make drone optimized and non-terrestrial networks (NTN) such as satellite connectivity a commonplace feature. For the Industrial Internet of things (IIoT) ecosystem, the release will offer connections from low-cost and low-data rate to extremely low latency with pinpoint accuracy (Lin, 2022).

In previous generations, the end-to-end network provided the same service to all users and the only option to offer guaranteed provision for a critical application, e.g., for public safety or critical infrastructure services was to deploy a dedicated physical network. In the 5GC network, network slicing allows operators to create thousands of virtual, independent networks within the same physical network infrastructure that connect from the device through to the application. Network slicing

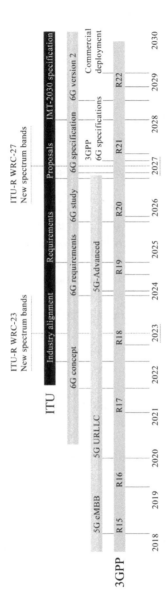

Fig. 2.1 Expected 3GPP standardization timeline and ITU-R process for IMT systems

enables operators to efficiently package novel 5G network capabilities into differentiated, guaranteed service level agreement-based (SLA) services in a cost-effective way.

The 5GA platform is visioned to introduce and extend a variety of novel applications and use cases across industries in 2025, and beyond (see, e.g., Ghosh et al., 2019; Nakamura, 2020).

- Extended mobile reality and ubiquitously available cloud gaming requires compact power-efficient devices supported by time critical communication capabilities.
- Wearable technology and devices demand a small form factor, efficiency, and high battery life.
- Industrial process monitoring and quality control are based on a massive volume of small data that should be transmitted frequently and efficiently to support network performance.
- Critical infrastructures such as public safety, railways and utilities with ultra-reliable low-latency communication combined with security and privacy requirements.
- Asset tracing and tracking in logistics demands extremely low energy consumption.
- Tele-operation of autonomous vehicles, robots, and drones demands reliable and secure communications both for the control and payload data.
- Location applications of connected devices with centimeter-level accuracy is enabled by advanced indoor and outdoor positioning technologies.
- Resilient, deterministic, and more stringent timing of 5G networks will be made affordably available and leveraged, e.g., in industrial automation to real-time financial transactions.

For 6G, ITU-R is working to publish the global framework for IMT toward 2030 and beyond in 2023 that will provide the basis for defining the future 6G.

Standardization

The worldwide success of mobile communications from the first generation onward can be seen to be largely founded on the initially proprietary technologies that have subsequently been transferred into a series of standards. Each new technology generation has required a decade of billions of euros investment in research and development to formalize technological innovations into standards and further into hardware and software products and services. Technology standardization has helped to generate foundational innovation platforms upon which emerging technology vendors have developed their products and services. From 1G onward, a similar standard release process has been followed providing standard blueprints for stakeholders to contribute and develop products and network solutions. The stakeholder community for the development has been well defined and stable consisting of a limited number of technology vendors, mobile network operators, system integrators, as well as academia and regulators.

With 5G, the technology ecosystem has been expanded particularly toward enterprises and industries introducing an unprecedented number of use cases and related novel stakeholder groups. Moreover, it should be acknowledged that 5G standardization deviates from previous generations having a coordinated single worldwide major approach to the IMT-2020 requirements. 3G (IMT-2000) was defined by three alternative paths (3GPP UMTS, 3GPP2 CDMA2000 and IEEE mobile WIMAX) and 4G (IMT-Advanced) with 3GPP LTE and IEEE mobile WIMAX alternatives that initially did not have an obvious single winner. Furthermore, 5G service-based architecture with open interfaces, the convergence of communication, information technology and data (ICDT), and user developer centricity will challenge the establish 3GPP grounded IMT process. Recent geopolitical and societal changes—especially related to discussions on data colonialization, user rights, and the use of artificial intelligence, and the digitalization of society and critical infrastructures—have given rise to discussions on the role of nations in standardization. The ongoing technology battle has specifically concerned the leadership in 5G regarding semiconductors, and concerns over sovereignty regarding AI and digital technologies have become an issue (Moerel & Timmers, 2021). As a recent example, the US "Clean network initiative" in 2020 addressed the long-term threat to data privacy, security, human rights, and principled collaboration to free the world from authoritarian malign actors (US Government, 2020). These developments raise the question of the possible fragmentation of the 6G standardization.

Role of Patents and Licensing

For a half-century, all major mobile communications technology providers have relied on patent licensing as their main value capture mechanism. The European telecommunications standards institute (ETSI) has orchestrated the development and governance of standards, controlling the technology contributors to make licenses available on a fair, reasonable, and non-discriminatory (FRAND) basis for a wide variety of implementers globally. The unique combination of technology co-development and widespread global adoption have been enabled by a nonexclusive licensing model. In addition to standard essential patent (SEP) royalties which have created a continuous incentive for standard contributions, technology vendors have leveraged complementarities via adjacent intellectual property (Teece, 2019).

The collaborative approach has empowered a downstream innovation and a mobile technology and application ecosystem. The standards-compliant ecosystem comprises dedicated technology/chipset firms, infrastructure equipment providers, mobile network operators, device manufacturers, operating system software providers, application developers, and content providers. Many specialized technology firms and vertically integrated companies in the mobile communications industry increasingly engage with two or more roles. Contrary to the single company-owned web-scale "winner-takes-all" digital platforms, harmonized common standards in mobile communications have helped define platforms with many stacked software layers.

A detailed look at the ETSI IPR online database (ETSI IPR) reveals that most 5G patents were declared between 2017 and 2019, and 25% of them were evolutionarily declared already for 4G. The database indicates that radio access networks (RAN), comprising the radio performance, physical layer, radio resource management specification, specification of the access network interfaces, the definition of the operations, and management requirements and conformance testing for user equipment and base stations encompasses about 84% of the SEPs. Physical layers 1 and layer 2 alone add up to 70% of SEPs. Services and systems aspects (SA) covering the overall architecture encompasses approximately only 11% of SEPs despite their leading role in security, management, orchestration, charging, and mission-critical applications areas. The remaining approximately 5% of SEPs are encompassed in the core network and terminals (CT) domain where differentiation and user experience have traditionally been implemented via technology system integration and overall network design, management, and orchestration. All in all, what matters is the device relevance found to be 80–90% of all the SEPs, which is in line with the distribution of licensing royalties (Yrjölä et al., 2022).

With a massive diffusion into new application areas and expanding the circle of stakeholders and licensees in the 6G era, firms may increasingly cooperate vertically in open dynamic multi-layered architectures while competing horizontally to capture value across services. The resulting complex licensing landscape will necessitate more precise rules for FRAND licensing as the exact interpretation and the associated reasonable licensing fees are not precisely defined in the current model (Teece, 2019). The extension toward cross-layered architecture functionalities and including data and algorithms will lead to the convergence of multiple connected ecosystems, introducing new roles and actors, especially related to system integration, management, and orchestration (Yrjölä et al., 2021).

Flexibility, scalability, and efficiency requirements combined with the long-tailed distribution needs of applications, may lead the 6G system to only specify a few core capabilities for the lower system layers with related open interfaces. Thus, higher layer distinct use case specifications for a complete connectivity platform will be done by different actors. For scalability and replicability among connectivity services, the lower-layer processing-intensive radio functions may continue to be specified by global standardization and continued to be implemented in custom silicon chipsets. On the other hand, the modular architecture with open interface specifications will enable the rest of the softwarized, programmable, and virtualized functions to be deployed on any commercial computing hardware. This will facilitate competition and entry, enabling stakeholders to access complementary assets through various forms of alliance with larger firms as well as to specialize within the ecosystem and develop complements to the platform. This suggests that value should be captured increasingly across multiple protocol layers and levels of the industry, and that the role of the de facto standard will need to be revised. Standards for systemic and complex general-purpose technologies, as Fig. 2.2 summarizes, will require coopetitive (i.e., simultaneous collaboration and competition) development to gain interoperability across ecosystems and industries.

One of the key challenges related to profiting from technological innovations in the 6G era is the protection and enforcement of intellectual property while fostering wide diffusion in the ecosystem. For example, starting from the discussion about who should acquire and pay for an SEP license: the OEMs, end-product manufacturers, or connectivity and application module suppliers. It will be essential to find a ruling that avoids the courts' protracted resolution of licensing disputes, ensures adequate compensation for developers, and promotes widespread use of innovations through appropriate fees. A compromised ETSI FRAND model and a more proprietary vertically integrated model with the reduced IP protection may be priced into products and services (Teece, 2019) and severely reduce the existing significant positive externalities that mobile communications technologies offer and place the envisioned 6G role as a general-purpose technology at risk.

Regulatory Developments

The mobile communications sector is tightly regulated. Regulation takes place at national, regional, and international levels via different methods and focus areas. One fundamental area is spectrum regulation, because the radio spectrum is the most critical natural resource needed for all wireless communications (Anker, 2017). Mobile communication

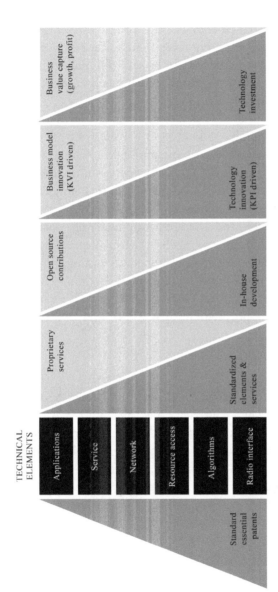

Fig. 2.2 The transformation of technology innovation across 6G system architecture layers stems from the 3rd Generation Partnership Project (3GPP) system architecture (Adapted from Yrjölä et al., 2022)

networks need spectrum to operate on and so do all other wireless communication systems such as satellites and terrestrial broadcasting, among others. However, if they use the same spectrum resources, there can be harmful interference that leads to significant service degradations. As a result, different wireless systems have traditionally sought their own exclusive use of the radio spectrum, which has been the foundation for mobile communications.

At the global level, the ITU-R sets requirements for systems to become part of the IMT family, that currently comprises 3G, 4G and 5G systems. At the regional level, coordination takes place between countries in specific regional organizations. In Europe, countries belonging to the European Union follow the European electronic communications code (EECC) directive, which defines the rules for electronic communication networks and services, and the spectrum used for mobile communications is harmonized. Many regulatory topics are a national matter including the actual spectrum awards determining who can deploy mobile communications networks and how. National level regulations consider international and regional approaches and define regulations that are considered appropriate in the country in question.

Spectrum Regulation

Spectrum regulation in particular plays a fundamental role in defining how, where, and when the developed technology is used and for what purpose (Matinmikko et al., 2014). Spectrum decisions made at the international, regional, and national levels significantly impact the resulting markets and the mobile communication sector is no exception. For mobile communications, every new technology generation has secured access to new spectrum, which has been internationally harmonized, leading to economies of scale by using the same equipment in larger markets.

Market regulations aim to achieve competitive markets where more than one MNO serves the end user customers in a country. Markets are directly impacted by spectrum regulatory decisions, especially via the rules in awarding of licenses. These national spectrum awarding decisions, which typically use spectrum auctions for mobile communications, significantly influence how many MNOs can operate in a country and how competitive the market is. Additionally, access regulation with rights and obligations concerning interconnection has a major influence on the markets.

Regulatory developments at the ITU-R regarding IMT-2000, IMT-Advanced and IMT-2020 systems have defined the development paths for 3G, 4G and 5G systems. The phases of regulatory development proceed from identifying technology trends and traffic characteristics to defining a joint vision, followed by detailed requirements definition, against which technology proposals are then evaluated. Finally, technology proposals that fulfill the requirements defined by the ITU-R become members of the IMT family and gain access to spectrum bands that are allocated to the mobile service and identified for IMT systems. The spectrum identification process goes in parallel with the IMT system process ranging from identifying spectrum needs based on market studies to studying candidate bands and their feasibility toward spectrum allocation decisions that are made at the World Radiocommunication Conferences (WRCs) of the ITU-R.

Regarding 6G, the process for IMT toward 2030 and beyond, which corresponds to 6G, is underway at the ITU-R. The technology trends have been identified and the report on future technology trends was published in 2022 (ITU-R, 2022). Work on the framework recommendation is ongoing and is expected to be completed in June 2023, presenting new usage scenarios for 6G. After WRC-23, which could develop an agenda item for the 6G spectrum for the following WRC in 2027 (WRC-27), the actual requirements definition phase will start in 2024. The requirements and needed evaluation criteria and processes will be finalized by the end of 2026. Technology proposals on 6G are expected in 2027–2028 with decisions taking place in 2029.

Regarding 6G, standardization phase 1 will likely start from 2025, leading to the first 6G specification in 3GPP Release 21 by 2028 and followed by commercial deployments around 2030. Meanwhile 5G will be enhanced by 5G-Advanced, which will be key focus for 3GPP in Release 18 and19 onward and will power commercial public and private networks starting in 2025. 5G-Advanced will provide new 5G features and boost 5G capabilities in four dimensions: experience, extension, expansion, and operational excellence.

The Evolution of the Business of Mobile Communications

The mobile communications industry has for long been referred to as an ecosystem (Zhang & Liang, 2011). In the current 4G-dominated world that is transitioning toward 5G dominance, the ecosystem comprises hardware providers, software providers, mobile equipment and infrastructure providers, content and application providers, network operators, content providers, OTT (over-the-top) Internet players, service providers such as MNOs (mobile network operators) and MVNOs (mobile virtual network operators), network infrastructure constructors, facility owners, regulatory bodies, and end users (Pujol et al., 2016). However, the way the ecosystem has been seen has changed over the history of mobile communication generations.

Latva-aho and Leppänen (2019) listed 29 different stakeholders for the envisioned 6G ecosystem, categorizing them into human, machine, enterprise, and public-sector type users, each with different demands and needs. In addition, they divided the stakeholders to have two different roles. *Resource and asset stakeholders* comprise device suppliers, network/cloud infra vendors, complementary technology providers, national regulators, public sector, government, data owners, context providers, content providers, context owners, edge cloud, data center, facility owner, site supplier, and building constructors. Meanwhile, *matching and bridging stakeholders* included mobile virtual network operators, mobile network operators, fixed operators, satellite operators, vertical-specific service providers, roaming service providers, application providers, digital twin providers, management service providers, data brokers, network resource brokers, broking/bridging providers, trust providers, and providers of security as a service.

From Value Chain to Business Ecosystem

Thus, the value chain in the mobile communication sector has evolved over the technology generations. The 2G era included state-owned MNOs and the market was opened to competition from new private MNOs. The value chain in 2G typically consisted of network infrastructure vendors, MNOs, device vendors, end users, and the regulator as depicted in Fig. 2.3. 3G introduced mobile broadband, which made new services and applications available over the networks. Otherwise, the value chain remained as it was in 2G, but competition increased in several markets with new market entry, leading to market consolidation later. 4G brought mobile broadband on a large scale and MNO networks became bit pipes for OTT services. In the 4G era, the role of OTT services increased and the number of MNOs per country decreased as a result of acquisitions by the MNOs.

The 5G era has introduced local networks deployed by different stakeholders, which has opened the market for new local entry. This development is still ongoing and varies a great deal between countries (Matinmikko et al., 2018). Local 5G networks have created local vertical specific ecosystems around their deployment areas where the stakeholders and their roles vary. Examples of this include the port and factory ecosystems.

From Engineering Platform to Service Modularity and Ecosystem

The definition of 5G opened the opportunity to change from connectivity-centric business models toward various connectivity with bundled content (data-based), context (location-based or service-specific), and commerce (platform) business models and offering the whole network as a service (NaaS). In parallel to this development, a disruption in the deployment models of mobile communication networks took place in the 5G era disrupting the ecosystem by enabling new entrants, such as utilities, ports, and manufacturing plants, to run their own local private

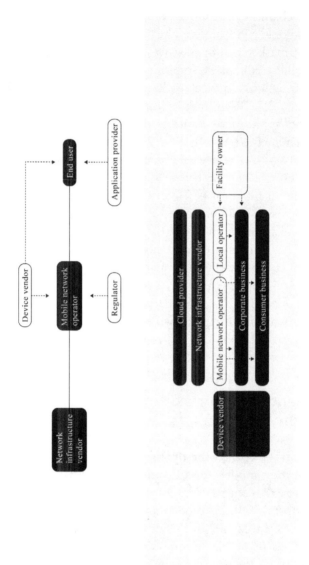

Fig. 2.3 The value chain in 2G, 3G, 4G (upper part of the figure), and 5G (lower part of the figure)

5G networks (Matinmikko et al., 2017). Additionally, other technologies such as cloud computing, AI, and Web3 have started to converge with or complement 5G introducing cloud computing "the fifth C" into the 4C business model characterization framework (Wirtz et al., 2010) as depicted in Fig. 2.4.

Figure 2.4 illustrates the evolutionary view of the mobile communications system from the 4G engineering connectivity platform via 5G

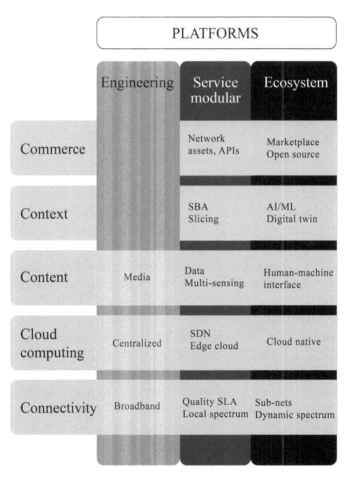

Fig. 2.4 From engineering platforms toward service modularity and ecosystem platforms

service modularity toward 6G connectivity platform-based ecosystem. In the 5C framework, the *connection* layer includes physical and virtualized communication network infrastructures for the ecosystemic value proposition of exchanging information. The newly introduced *cloud computing* infrastructure is an essential enabler for a variety of data and intelligence-based services. The third *content* layer aims to collect, select, compile, distribute, and present data in the ecosystem in a value-adding, convenient, and user-friendly way. In the *context* layer, the aim is to provide a structure, increase transparency, and reduce complexity by providing a platform for stakeholders' communication and transaction. Finally, the *commerce* layer focuses on negotiation, initiation, payment, and service and product deliveries in the ecosystem, enabling low transaction costs and providing a cost-effective marketplace for matching and bridging supply and demand.

Despite massive investment in the current mobile communication networks, the MNOs' opportunities for differentiation have been limited. The differentiation capacity has shifted toward devices and content and the mobile operating systems have become bottleneck assets in the mobile ecosystem. 5G can be seen as a service modular platform system stemming from interfaces that enable complementary offerings of elements and services. The digital platform business model enables software developers to add value through applications and complementary assets to the ecosystem by attracting users and building network effects. The ecosystemic 6G connectivity platform-based model facilitates value co-creation, co-capturing, and sharing to maximize the overall value generated and acquired not only by a focal traditional incumbent MNO but also by the ecosystem's stakeholders. The 6G ecosystem can be seen both as an innovation engineering platform and *e-commerce* transaction platform (Evans & Gawer, 2016). This will enable digital business ecosystems to facilitate exchanges of otherwise fragmented groups of consumers and/or firms and to provide a technology and distribution system for other companies to base their technological and service innovations. Introduced service modularity on 5G platforms will on one hand enable fast-paced autonomous innovation, but on the other hand it will change the appropriability mechanism by reducing the role of complementary core assets.

Technological Architecture as a Starting Point

In 6G, the systemic architecture level innovations will be vital in enabling business model changes. Key transformational bottleneck assets such as AI/ML and human–machine interfaces (HMI) as general-purpose technologies will be leveraged across distributed a heterogeneous 6G cloud architecture. Intelligent 6G networks are based on common enablers for AI as a service, and federated learning as-a-service that leverage data acquisition, data exposure and a common cross-domain analytics framework. Intelligent network enablers will operate across the cloud continuum from the central cloud to the edge and to far edge including the UE. Extreme scalability and flexibility will become the new paradigm in 6G. Network automation and orchestration will be integral parts of intelligent networks using AI and analytics to manage and orchestrate the networks in a fully automated manner across all layers and parts of the network abstraction (Kaloxylos et al., 2021).

The primary focus in the current 4G and early 5G deployments has been on network planning, network diagnostics, and network optimization and control reducing capital expenditure, optimizing network performance, and building new revenue streams through the improved customer experience. 6G radios are envisioned to adopt AI/ML in a fundamental way for optimized air interface design, cognitive dynamic spectrum use, and context awareness. On the network level, hyper-specialized agile slicing will call for new fully AI automatized service management and orchestration for network automation, allowing dynamic adaptation of network resources according to changing service requests, reducing the deployment time of new services and mitigation of failures, and significantly reducing operating expenditure. Digital trust, enabled by quantum computing and distributed Web3 ledger technologies such as blockchain and smart contracts, will provide businesses securely and predictably with world-class cybersecurity, public safety, and fintech solutions.

Human augmentation will enable people to interact with and within the digital world. This will include VR headsets, XR glasses, remote control with haptics, and, in the future brain-machine interfaces and connected bio-medical implants. The fusion between the digital and physical realms will further enhance our capabilities to interact with dynamic representations of real-world objects, systems, and processes in the digital world such as digital twins and 6G network sensing data. Downstream digital application platforms will converge and there will be multimodal engagement with media, and the physicality of lived experiences will be seamlessly accessible through a HMI extended to all five senses, including the senses of touch and taste. Individual and collaborative users will seamlessly be able to switch between any form of immersive mobile extended reality, encompassing virtual reality, augmented reality, and mixed reality, comprising both virtual and augmented objects. HMI opportunities will be clearly differentiated between the consumer, the enterprise, and the industrial segment.

Toward the 6G World

Disruptions on multiple levels are a visible part of both organizational life as well as economic reality these days (e.g., Buckley, 2019). In the last few years, global industries have faced disruptions in the form of the China-US trade war, the technological war between different centers of power especially in emergent industries (Chin, 2019; Lukin, 2019; Petricevic & Teece, 2019), COVID-19, and more recently the Russia-Ukraine war. As a result, in recent times, we have witnessed a plethora of terms emerging; out of which the most famous is VUCA (volatility, uncertainty, complexity, and ambiguity) as a permanent feature of the current economy (Bennett & Lemoine, 2014; Millar et al., 2018) especially in industries which are significantly intertwined with global value creation. In this context, mobile communications is one of the sectors that has been very visibly linked to most elements of global disruptions because of its criticality to the economic competitiveness as well as its visible interlinkage to the emergent digital business models (Kilkki et al., 2018). A well-known example in this regard is the Chinese telecom giant

Huawei which has received bans and strict oversight in different Western countries including USA due to the concerns about privacy and security. As the shift toward 6G is taking shape globally, there is a race for setting the standards, and geopoliticalGeopolitical disruptions (and considerations) are a core aspect of this debate (Klement, 2021; Yrjölä et al., 2020). At the same time, it is vital to stress that the influence of a variety of global disruptions on 6G development and planned implementation has not been studied specifically so far; thereby showing a visible gap in the extant literature.

A Business Architecture Perspective

In the engineering tradition, platforms have been seen as modular technological designs for facilitating innovation, whereas in the economics tradition platforms have been seen as two- or multi-sided markets connecting supply and demand. The traditional approach to a mobile communications ecosystem is based on a layered protocol-based technical infrastructure, an engineering platform consisting of elements and interfaces. As in digitalized industries in general, in mobile communications, business models can be seen to follow the nature of integration—vertical or horizontal (Ballon, 2007). The previous 4G and 5G business architectures have considered the ecosystem configuration either through vertical or horizontal business models as depicted in Fig. 2.5.

In the *vertical business model*, traditionally employed in mobile communications, a firm controls its suppliers, distributors, or retail locations as part of its supply chain. To be competitive in this supply-sided model, a firm focuses on creating value for its customers, and is grounded inside its selected verticals. On the 4G engineering platform, a competitive advantage arises from focusing on value creation within narrow segments around connectivity and content (Ahokangas et al., 2019). As interfaces in mobile communications have been largely defined from the inside-out perspective, the telco APIs have not reached the developers' ecosystem, and infrastructure providers have been controlling the complete technology and service solution (Basole & Karla,

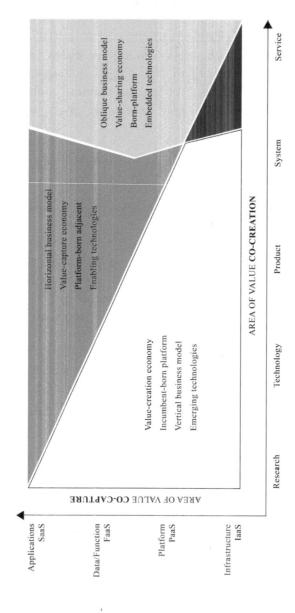

Fig. 2.5 Evolution from vertical and horizontal business logics toward oblique value creation and value capture in 6G (Adapted form Yrjölä et al., 2022)

2011). In the industry transition from 4G to 5G, the vertical integration strategy has been clearly visible with the acquisition of business operations within the same vertical. Deployed incumbent-born mobile communications connectivity platforms have typically been upstream platforms and dependent on the core product (Pundziene et al., 2022) and slow to respond to market dynamics.

The introduction of 5G has transformed the traditional vertical business model approaches of the mobile sector toward a horizontal model (Cave, 2018). The *horizontal business model*, adopted in consumer service-oriented businesses, to serve a wide clientele across different segments focuses on economies of scale and scope in order to maximize the value capture. This demand-side approach enables multiple stakeholders to focus on their respective fields through a common framework that allows faster innovations and a rapid scale-up of applications and businesses. Technological innovations, extreme cost consciousness, and risk awareness have been characteristic in capturing customer value while defending a position against competition. The horizontal 5G business models are highly dependent on the supporting infrastructure and complementors to run smoothly. The introduced service-based architecture (SBA) with softwarization and cloudification technology has enabled demand-side platformization (Camps-Aragó et al., 2019) that enables innovative as-a-service business models to serve a wider value constellation (Hmoud et al., 2020). The novel serverless cloud-native model allows developers to build and run applications without having to manage servers. IT and cloud Webscalers' platform-born adjacent platforms can serve downstream users via transformative service innovations (Pundziene et al., 2022).

In the 6G era, the vertical integration in the value chain and/or horizontal diversification to any segment is unlikely as digital service chains are becoming more distributed, abstracted, and advanced built on resources provided as-a-service. Concepts centered around network-as-a-service (NaaS) will be very mature by 2030, and everything that can be offered as a service will be (Yrjölä et al., 2022). Cloud native design, open source, and standards will drive openness in the architecture of networks and operations, while enabling technologies such as hyper-specialized virtualization and slicing, abstracted data, while analytics capabilities will provide the right building blocks (Yrjölä et al., 2021). Simple-to-consume low-code/no-code APIs and network-as-a-code, backed by new levels of AI/ML-driven automation will be the enablers of new business and revenue for all players. This will require increased coopetition among network function vendors, network service providers, application service providers, and hyperscale companies, within an evolved ecosystem. In modern IT and software development, DevOps and infrastructure-as-a-code (IaC) are already mainstream, and SW developers are the drivers of a new kind of innovation and service delivery (Morris, 2016). This multi-sided platform-of-platforms model integrates the supply and demand side and can be seen to form a sharing economy. Wide adoption and maturity of business-to-business marketplaces are emerging for enterprises and IT in hyperscale cloud ecosystems. These developments will define the traction for telco exposure and abstraction with mobile in moving toward hybrid oblique business models. The *oblique business model* views 6G as a general-purpose technology and envisages network-as-a-code for developers. A loosely coupled oblique business model (Amit & Zott, 2015; Saebi & Foss, 2015) can be seen to follow the rationales of open innovation (Chesbrough, et al., 2014) and the timely concept of a sharing economy in which resource efficiency plays a crucial role (Stephany, 2015). In the 6G era, business models will not be built on one-sided technology or in industry silos, because it will be essential to consider the lifecycle stage of complementors, customers, and partners in the ecosystem. In a value-sharing economy, a cumulative, open-sourced effort of a community of developers will turn customers', and ecosystems' underutilized assets into more efficient or better used assets with fast-to-market strategies (Bogers & West, 2012; Chesbrough et al., 2014). Stakeholder interactions will aim to achieve common strategic objectives and eventually share a common fate and will no longer be based on customer–supplier relationships, (Iansiti & Levien, 2004). The oblique 6G business model characteristics will enable a novel born-platform approach, which will be a stand-alone multi-sided platform type of architecture building on a digital platform value proposition from the beginning of a new venture aiming at new market creation (Nambisan, 2017; Pundziene et al., 2022).

Implications

Despite the wide streams of platform and ecosystem business literature, little effort has been made to advance a coherent theory on hybrid ecosystemic platform-based business models that combines the characteristics of the both the transaction platforms focusing serving or mediating exchange and interactions (McIntyre & Srinivasan, 2017, p. 143), and the innovation platform creating value through enabling innovations on the platform (Cusumano et al., 2020). 6G platforms can be seen as a composition of interacting subsystems that will always to some degree be interdependent and interoperate exclusively using predefined, stable interfaces (Eisenmann et al., 2006).

We emphasize the role of multi-sided platforms (Teece et al., 2022) to organize collaboration and control without owning the services whose exchange it inter-organizationally facilitates and governs. Moreover, the value co-creation (Saebi & Foss, 2015) by a compilation of peripheral enterprises connected to the platform via shared or open-source technologies or technology standardization (Cennamo & Santaló, 2013; Jacobides et al., 2018) will be seen in 6G. Similarly, the logic of value co-captured via sharing and distributing the revenue among ecosystem members, not only to capture by the focal enterprise can be envisioned (Oh et al., 2015; Upward & Jones, 2016; Zott et al., 2011).

The 6G ecosystem will be associated with both product and service systems (Tsujimoto et al., 2018), and both upstream and downstream value network actors, as well as related technologies and institutions with a varying degree of not fully hierarchically controlled multilateral, non-generic complementarities (Adner, 2017; Iansiti & Levien, 2004; Jacobides et al., 2018). This ecosystemic model will build upon customer-centricity (Weil & Woerner, 2015) and the dynamics of industry transformation, moving toward sustainable business, and can be viewed as an engine of social progress (Lüdeke-Freund et al., 2018), while the value-in-exchange may be captured on multiple levels (Lepak et al., 2007). The discussion above leads to the proposition of seeing ecosystem and platform concepts to be intertwined and thus defines a 6G ecosystemic platform-based business model as follows: *An ecosystemic platform-based business model will connect various sides of multi-sided markets to facilitate value co-creation, co-capture, and sharing on multiple levels over a platform that facilitates interaction between users to collectively create innovations* via *matching, complementing, or sharing their resources sustainably.*

Bibliography

Adner, R. (2017). Ecosystem as structure: An actionable construct for strategy. *Journal of Management, 43*(1), 39–58.

Ahokangas, P., Atkova, I., Yrjölä, S., & Matinmikko-Blue, M. (2023). Business model theory and the becoming of mobile communications technologies. In A. Aagaard, & C. Nielsen (Eds.), *BMI game changers.* Palgrave MacMillan.

Ahokangas, P., Matinmikko, M., Yrjola, S., Okkonen, H., & Casey, T. (2013). Simple rules for mobile network operators' strategic choices in future cognitive spectrum sharing networks. *IEEE Wireless Communications Magazine, 20,* 20–26.

Ahokangas, P., Matinmikko-Blue, M., Yrjölä, S., Seppänen, V., Hämmäinen, H., Jurva, R., & Latva-aho, M. (2019). Business models for local 5G micro operators. *IEEE Transactions on Cognitive Communications and Networking, 5*(3), 730–740.

Amit, R., & Zott, C. (2015). Crafting business architecture: The antecedents of business model design. *Strategic Entrepreneurship Journal, 9*(4), 331–350.

Anker, P. (2017). From spectrum management to spectrum governance. *Telecommunications Policy, 41*(5–6), 486–497.

Ballon, B. (2007). Business modelling revisited: The configuration of control and value. *Info, 9*(5), 6–19.

Basole, R. C., & Karla, J. (2011). On the evolution of mobile platform ecosystem structure and strategy. *Business & Information Systems Engineering, 3*(5), 313.

Bennett, N., & Lemoine, G. J. (2014). What a difference a word makes: Understanding threats to performance in a VUCA world. *Business Horizons, 57*(3), 311–317.

Bogers, M., & West, J. (2012). Managing distributed innovation: Strategic utilization of open and user innovation. *Creativity and Innovation Management, 21*(1), 61–75.

Buckley, P. (2019). *International business in a VUCA world: The changing role of states and firms*. Emerald Group Publishing.

Cave, M. (2018). How disruptive is 5G? *Telecommunications Policy, 42*(8), 653–658.

Camps-Aragó, P., Delaere, S., & Ballon, P. (2019). 5G business models: Evolving mobile network operator roles in new ecosystems. *Smart Cities & Information and Communication Technology (CTTE-FITCE)*, 1–6.

Cennamo, C., & Santaló, J. (2013). Platform competition: Strategic trade, tradeoffs in platform markets. *Strategic Management Journal, 34*(11), 1331–1350.

Chen, W., Montojo, J., Lee, J., Shafi, M., & Kim, Y. (2022). The standardization of 5G-Advanced in 3GPP. *IEEE Communications Magazine, 60*(11), 98–104.

Chesbrough, H., Vanhaverbeke, W., & West, J. (2014). *New frontiers in open innovation*. Oxford University Press.

Chin, W. (2019). Technology, war and the state: Past, present and future. *International Affairs, 95*(4), 765–783.

Cusumano, M. A., Gawer, A., & Yoffie, D. B. (2019). *The business of platforms: Strategy in the age of digital competition, innovation, and power*. Harper Business.

Cusumano, M. A., Yoffie, D. B., & Gawer, A. (2020). The future of platforms. *MIT Sloan Management Review, 61*(1), 26–34.

David, K., & Berndt, H. (2018). 6G vision and requirements: Is there any need for beyond 5G? *IEEE Vehicular Technology Magazine, 13*(3), 72–80.

Eisenmann, T., Parker, G., & Van Alstyne, M. W. (2006). Strategies for two-sided markets. *Harvard Business Review, 84*(10), 92.

Evans, P. C., & Gawer, A. (2016). *The rise of the platform enterprise: A global survey*. Center for Global Enterprise.

Ghosh, A., Maeder, A., Baker, M., & Chandramouli, D. (2019). 5G evolution: A view on 5G cellular technology beyond 3GPP release 15. *IEEE Access, 7*, 127639–127651.

Hmoud, A. Y., Salim, J., & Yaakub, M. R. (2020). Platformisation of mobile operators business model: A proposition using design science approach and grounded theory principles. *International Journal on Advanced Science Engineering Information Technology, 10*(2), 473–484.

Iansiti, M., & Levien, R. (2004). *The keystone advantage: What the new dynamics of business ecosystems mean for strategy, innovation, and sustainability*. Harvard Business Press.

ITU-R. (2020). IMT Vision—Framework and overall objectives of the future development of IMT for 2020 and beyond. International Telecommunication Union Radiocommunication Sector. https://www.itu.int/rec/R-REC-M.2083-0-201509-I.

ITU-R. (2022). *Future technology trends of terrestrial International Mobile Telecommunications systems towards 2030 and beyond*. International Telecommunication Union Radiocommunication Sector. https://www.itu.int/pub/R-REP-M.2516.

Jacobides, M. G., Cennamo, C., & Gawer, A. (2018). Towards a theory of ecosystems. *Strategic Management Journal, 39*(8), 2255–2276.

Kaloxylos, A., Gavras, A., Camps, D., Ghoraishi, M., & Hrasnica, H. (2021). *AI and ML–Enablers for beyond 5G Networks*. https://www.recercat.cat/bitstream/handle/2072/522533/AI-MLforNetworks-v1-0.pdf?sequence=1.

Kilkki, K., Mäntylä, M., Karhu, K., Hämmäinen, H., & Ailisto, H. (2018). A disruption framework. *Technological Forecasting and Social Change, 129*, 275–284.

Klement, J. (2021). *Geo-economics: The interplay between geopolitics, economics, and investments*. https://www.cfainstitute.org/-/media/documents/book/rf-publication/2021/geo-economics-full.pdf.

Latva-aho, M., & Leppänen, K. (2019). *Key Drivers and Research Challenges for 6G Ubiquitous Wireless Intelligence* [White paper]. University of Oulu. https://www.6gflagship.com/key-drivers-and-research-challenges-for-6g-ubiquitous-wireless-intelligence/.

Lepak, D. P., Smith, K. G., & Taylor, M. S. (2007). Value creation and value capture: A multilevel perspective. *Academy of Management Review, 32*(1), 180–194.

Lin, X. (2022). An overview of 5G advanced evolution in 3GPP release 18. *IEEE Communications Standards Magazine, 6*(3), 77–83.

Lüdeke-Freund, F., Carroux, S., Joyce, A., Massa, L., & Breuer, H. (2018). The sustainable business model pattern taxonomy—45 patterns to support sustainability-oriented business model innovation. *Sustainable Production and Consumption, 15*, 145–162.

Lukin, A. (2019). The US–China trade war and China's strategic future. *Survival, 61*(1), 23–50.

Matinmikko, M., Mustonen, M., Roberson, D., Paavola, J., Höyhtyä, M., Yrjölä, S., & Röning, J. (2014). Overview and comparison of recent spectrum sharing approaches in regulation and research: From opportunistic unlicensed access towards licensed shared access. *IEEE DySPAN, McLean, VA, USA, 2014*, 92–102.

Matinmikko, M., Latva-aho, M., Ahokangas, P., Yrjölä, S., & Koivumäki, T. (2017). Micro operators to boost local service delivery in 5G. *Wireless Personal Communications, 95*, 69–82.

Matinmikko, M., Latva-aho, M., Ahokangas, P., & Seppänen, V. (2018). On regulations for 5G: Micro licensing for locally operated networks. *Telecommunications Policy, 42*(8), 622–635.

McIntyre, D. P., & Srinivasan, A. (2017). Networks, platforms, and strategy: Emerging views and next steps. *Strategic Management Journal, 38*(1), 141–160.

Millar, C. C., Groth, O., & Mahon, J. F. (2018). Management innovation in a VUCA world: Challenges and recommendations. *California Management Review, 61*(1), 5–14.

Moerel, L., & Timmers, P. (2021). *Reflections on digital sovereignty.* https://eucd.s3.eu-central-1.amazonaws.com/eucd/assets/khGGovSY/rif_timmersmoerel-final-for-publication.pdf.

Morris, K. (2016). *Infrastructure as code: managing servers in the cloud.* O'Reilly Media, Inc.

Nakamura, T. (2020, June). 5G Evolution and 6G. In *2020 IEEE symposium on VLSI technology* (pp. 1–5). IEEE.

Nambisan, S. (2017). Digital entrepreneurship: Toward a digital technology perspective of entrepreneurship. *Entrepreneurship Theory and Practice, 41*(6), 1029–1055.

Oh, J., Koh, B., & Raghunathan, S. (2015). Value appropriation between the platform provider and app developers in mobile platform mediated networks. *Journal of Information Technology, 30*(3), 245–259.

Parkvall, S., Dahlman, E., Furuskar, A., & Frenne, M. (2017). NR: The new 5G radio access technology. *IEEE Communications Standards Magazine, 1*(4), 24–30.

Petricevic, O., & Teece, D. J. (2019). The structural reshaping of globalization: Implications for strategic sectors, profiting from innovation, and the multinational enterprise. *Journal of International Business Studies, 50*(9), 1487–1512.

Pujol, F., Elayoubi, S. E., Markendahl, J., & Salahaldin, L. (2016). Mobile telecommunications ecosystem evolutions with 5G. *Communications & Strategies, 102*, 109.

Pundziene, A., Gutmann, T., Schlichtner, M., & Teece, D. J. (2022). Value impedance and dynamic capabilities: The case of medtech incumbent-born digital healthcare platforms. *California Management Review, 64*(4), 108–134.

Saebi, T., & Foss, N. J. (2015). Business models for open innovation: Matching heterogenous open innovation strategies with business model dimensions. *European Management Journal, 33*(3), 201–213.

Stephany, A. (2015). *The business of sharing: Making it in the new sharing economy.* Palgrave MacMillan.

Teece, D. J. (2019). *5G and the global economy: How static competition policy frameworks can defeat open innovation.* https://static1.squarespace.com/sta tic/5a83233eb0786912d9c89545/t/5d9cbc3217af3e320d97270a/157055 2889394/CPI_Antitrust_5GGlobalEcon_Sept2019.pdf.

Teece, D. J., Pundziene, A., Heaton, S., & Vadi, M. (2022). Managing multi-sided platforms: Platform origins and go-to-market strategy. *California Management Review, 64*(4), 5–19.

Tsujimoto, M., Kajikawa, Y., Tomita, J., & Matsumoto, Y. (2018). A review of the ecosystem concept—towards coherent ecosystem design. *Technological Forecasting and Social Change, 136*, 49–58.

The United States government (2020). *The Clean Network.* https://2017-2021. state.gov/the-clean-network/index.html.

Upward, A., & Jones, P. (2016). An ontology for strongly sustainable business models: Defining an enterprise framework compatible with natural and social science. *Organization & Environment, 29*(1), 97–123.

Weill, P., & Woerner, S. L. (2015). Thriving in an increasingly digital ecosystem. *MIT Sloan Management Review, 56*(4), 27.

Wirtz, B.W., Schilke, O., & Ullrich, S. (2010). Strategic development of business models: Implications of the web 2.0 for creating value on the internet. *Long Range Planning, 43*(2–3), 272–290.

Yrjölä, S., Ahokangas, P., & Matinmikko-Blue, M. (2021). Platform-based business models in future mobile operator business. *Journal of Business Models, 9*(4), 67–93.

Yrjölä, S., Ahokangas, P., & Matinmikko-Blue, M. (2022). Value creation and capture from technology innovation in the 6G Era. *IEEE Access, 10*, 16299–16319.

Zhang, J., & Liang, X. J. (2011). Business ecosystem strategies of mobile network operators in the 3G era: The case of China Mobile. *Telecommunications Policy, 35*(2), 156–171.

Zott, C., Amit, R., & Massa, L. (2011). The business model: Recent developments and future research. *Journal of Management, 37*(4), 1019–1042.

3

Future Scenarios and Anticipated Impacts of 6G

Seppo Yrjölä, Petri Ahokangas,
and Marja Matinmikko-Blue

By being unknowable, by resulting from events which, at the sub-atomic level, cannot be fully predicted, the future remains malleable, and retains the possibility of change, the hope of coming to prevail; victory, to use an unfashionable word. In this, the future is a game; time is one of the rules.

(Iain M. Banks. The player of games)

S. Yrjölä (✉)
Centre for Wireless Communications, University of Oulu, Oulu, Finland
e-mail: seppo.yrjola@oulu.fi; seppo.yrjola@nokia.com

Nokia, Oulu, Finland

P. Ahokangas
Martti Ahtisaari Institute, Oulu Business School, University of Oulu, Oulu, Finland
e-mail: petri.ahokangas@oulu.fi

© The Author(s) 2024   **45**
P. Ahokangas and A. Aagaard (eds.), *The Changing World of Mobile Communications*,
https://doi.org/10.1007/978-3-031-33191-6_3

Framing Change Toward 6G

Disruptions at multiple levels are a visible part of both organizational life (e.g., Greene et al., 2018; Ivanov et al., 2019) and economic reality (Buckley, 2019; Dieppe, 2021; Klement, 2021). In the last few years, global industries have faced disruptions in form of China-US trade war (e.g., Cho & Moon, 2022; Lukin, 2019), technological war between different centers of power especially in the emergent industries (e.g., Chin, 2019; Klement, 2021; Petricevic & Teece, 2019), COVID-19 (e.g., Ali et al., 2022; Ivanov, 2021), and more recently Russia-Ukraine war (Korn & Stemmler, 2022; OECD, 2022). As a result, in recent times, we have witnessed a plethora of terms emerging, out of which the most famous is VUCA (volatility, uncertainty, complexity, and ambiguity) as a permanent feature of the current economy (Bennett & Lemoine, 2014; Buckley, 2019; Greene et al., 2018; Millar et al., 2018) especially in the industries, which are significantly intertwined with global value (and supply) chains (Agarwal et al., 2021) such as 6G mobile communications. Studies focusing on VUCA strategies have stressed the critical role of agility to deal with volatility, information to deal with uncertainty, restructuring to deal with complexity, and experimentation to deal with ambiguity (Bennett & Lemoine, 2014; Buckley, 2019; Greene et al., 2018; Ozbayrac, 2022).

Both 5G and 6G are expected to bring disruptive changes to our increasingly digitalized world as these technologies have been considered to be the backbone for future digitalization. This chapter aims to provide a holistic and comprehensive framework depicting how 5G and 6G could be developed in the future—and what kind of impact these technologies could have when analyzed at multiple levels. The analysis presented in this chapter serves as a high-level starting point for

M. Matinmikko-Blue
Infotech Oulu Focus Institute and Centre for Wireless Communications, University of Oulu, Oulu, Finland
e-mail: marja.matinmikko@oulu.fi

the subsequent chapters that drill down into more detailed discussions relevant for the future development of mobile communications.

This chapter takes the form of an exploratory scenario analysis that follows a traditional process comprising the identification of key drivers and uncertainties and the generation of alternative scenarios (Schoemaker, 1995). This work is based on the data collected from the expert group workshops organized by the 6G Flagship (6G Flagship White Papers, 2020), where a total of 146 forces were discovered and identified, and 23 key drivers (trends and uncertainties) were selected based on their anticipated impact and predictability of consequences. Based on the key trends, 16 scenarios were generated and assessed for their probability, plausibility, and preferability. This chapter ends with a summary of the scenarios regarding their economic, societal, and environmental consequences.

Key Trends for the 5G/6G Transition

Several key trends for 6G were identified based on the ranking of forces by their high impact and high probability as shown in Fig. 3.1. They are further discussed in the following.

Political, Legal, and Regulatory

Public network funding has traditionally been directed at unserved and underserved areas in terms of broadband access and coverage. Recently, support for deployment programs has extended to areas such as smart city community development, logistic hubs (such as ports and airports), advanced health services, public safety, and critical infrastructure. Smart infrastructures are expected to be hyperconnected and completely automated. They will serve as a middle layer between humans and natural environments. These networks will be put together with a public–private-personal partnership and ownership funding model with a view to sustainable growth and the use of digital infrastructure.

Wireless communications spectrum politics and spectrum management in the 6G era will reveal a new level of complexity that stems

TRENDS

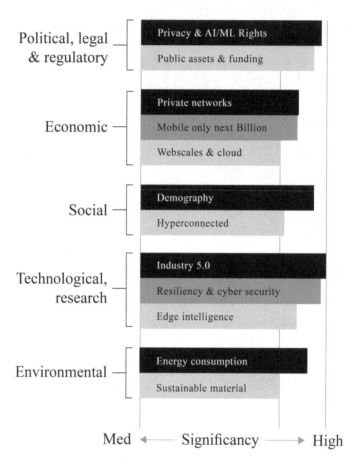

Fig. 3.1 Identified key trends

from the variety of spectrum bands and spectrum access models with different levels of sharing. *Local deployments* of networks by a variety of stakeholders are expected to grow further in 6G. The timescales of international-level spectrum management will no longer be enough with the rapid technological development of mobile communication networks and changing user needs. Spectrum sharing will play an increasing role in accommodating new 6G systems with existing spectrum users. Furthermore, in national technology politics, spectrum regulation will be used to gain a competitive advantage. Site densification of the mobile communication infrastructure is happening in the higher spectrum band deployments creating scarcity of real estate with different pay-back periods for information (IT) and telecommunication technologies (5–10 years) vs. infrastructure (15–30 years).

Telecom networks have become essential enablers of *National sovereignty* and critical infrastructure platforms on the level of electricity and water supply. This sets requirements for the redundancy and resilience of services, networks, infrastructure, and businesses. Furthermore, governments are active in the technology governance in 6G development and look to reassert control and technology leadership over key technologies.

Economic

Webscalers are increasingly dominating ecosystems and future IT and ICT market dynamics and will be gaining dominance in the telecom cloud infrastructure by acquiring assets and talent to obtain a bigger slice of the pie from edge cloud monetization. Over-the-top (OTT) companies will increasingly utilize their customer data, cloud infrastructure, and AI/ML capabilities to challenge traditional operators' customer relationship ownership, because users will be valuing service experience and perceive connectivity as a basic utility. In addition to the media space, OTT players will offer basic telco services such as voice or messaging and are expected to be active in growth areas such as cloud space and services, competing with telcos for clients and revenue. They will tie customers to their own ecosystems with carrier-neutral connectivity, while making reliance on traditional operators a thing of the past.

Social

In *Mobile-only next billion,* ubiquitous cheap user terminals and increasingly affordable network connections in megacities and rural areas will help another billion users join the Internet and increasingly access applications and digital content aimed at non-English-speaking markets. Currently, mobile is for many the primary or only channel for accessing the Internet and services. Non-terrestrial networks (NTN) for connectivity will be deployed in areas not covered by terrestrial networks (~95% of earth's surface). Huge investments from cash-rich companies and nations to establish LEO (low-earth-orbit) satellite mega constellations may provide attractive balance between global coverage, high capacity, low latency, and TCO with rapidly decreasing satellite production and launch costs. Furthermore, sustainable radios in NTN (which represent > 80% of power consumption in macro-terrestrial) will be solar powered in space. With its unprecedented scale and growing impact on daily lives, mobile is a powerful tool for achieving the United Nations' Sustainable Development Goals (SDGs) and for driving sustainable economic growth.

Private networks driven by industrial digital automation call for standalone networks with high reliability, high performance in terms of both bandwidth and reliability, secure communications, and data privacy, fulfilling business and mission-critical needs. The solutions will enable the integration of processes, data, and diverse devices such as sensors, machines, and in-vehicle and hand-held devices across a wide range of applications for industry enterprises. Private networks could be established without direct MNO involvement. Furthermore, demands for privacy in personal space may also create private subnetworks that rarely connect with the public Internet.

In 2015, 85% of global GDP was generated in cities. Urbanization will see 5 billion people living in cities by 2030, occupying 3% of the earth's land, but accounting for 80% of energy consumption and 75% of carbon emissions. Ninety-five percent of urban expansion in the next decades will occur in the developing world, where 883 million people live in slums today. Rapid urbanization is exerting pressure on freshwater supplies, sewage, the living environment, and public health. Future *demographics* divide a world of 8.6 billion inhabitants by 2030 into two camps: a growing one in sub-Saharan Africa and South Asia and a stalling and decreasing one, including Europe, Russia, and post-peak China. Future cities will be hungry global economic engines and the economic powerhouses of the global economy. Cities are increasingly functioning as autonomous entities, setting social and economic standards. Urban identity will grow in importance compared with national identity.

A hyperconnected globe will continue to feel ever smaller in 2030: globally, 90% will be able to read, access the Internet, and be on the move. 6G will transform urban and rural living, existing at the intersection of geopolitics, the growth of nationalism, rights to information transparency, and information citizenry. Thus, once the 6G infrastructure is in place, content growth will lie in supporting multiple social and technological identities of people through a variety of media. This will require a mindful view of decision making and the regulation of future data, information, media, and network usage in light of sustainable economic growth. Thus, the people in 6G worlds will be increasingly sophisticated in their media and service consumption, while being rooted in their local economies. Connectivity will therefore be not only virtual and digital, but physical. Furthermore, the approaching opportunity to redefine the *human–machine interface* (HMI) will enable the biological world to be connected in novel ways. Many future mobile devices will become thinner and lighter in a variety of form factors while offering massive computational capabilities at the same time for applications such as truly immersive XR, mobile holograms, and digital replicas. 6G will make it possible to eliminate wired communication entirely, at least for short-range transmission. Highly specialized radio subnetworks will be installed, where applications run in robots, production modules vehicles, and even human bodies. These autonomous specialized radio cells will be able to support life-critical services that cannot depend on connections to the overlay network, though they can benefit when those connections are available.

Technology and Research

Industry 5.0 (I5.0) will allow collaborative HMI between services and industries, because human intelligence will be in perfect harmony with advanced cognitive computing. With real-time data, effective data monetization, and digital automation of the manufacturing process, businesses will be able to shift the focus to generating higher revenues from the servitization of products. Advanced manufacturing capabilities will help to overcome design complexities with the ability to facilitate an extremely long tail of mass customization and further return control to customers in a haptic way. Furthermore, 15.0 will require the highest standards of safety and environmental protection.

The need for *resiliency, cybersecurity, and trust* will be ubiquitous in the hyperconnected world of 2030. Digitalization is permeating the biological and physical worlds, thereby making security a question of life and death. The number of threats is increasing significantly especially due to the billions of devices and subnetworks of different sources deployed, in addition to open interfaces and disaggregation-enabled microservices from multiple vendors and open sources which cannot be trusted. Even a temporary loss of technology may have not only a productivity but a psychological impact on our lives. Furthermore, the subversion or corruption of our technology may result in disastrous harm to our lives and businesses, for example, if medical treatment devices deliver the wrong medication, education systems teach propaganda, or work automation causes injury or damage to our products and businesses. In particular, expectations toward protecting and safeguarding society and critical infrastructure from emergency situations by means of technological advances are anticipated to grow. Quantum-safe encryption and a variety of physical-layer and architecture security mechanisms are emerging for trusted service delivery on a zero-trust infrastructure, which will become an integral part of the 6G network.

Growing amounts of data and requirements for low latency are taking computing to the *edge cloud* and decentralizing network topology in a distributed cloud configuration. With the growth of extreme edge intelligence, the proliferation of increasingly powerful communication, computing, and analytics resources at the edge of the network will convert architecturally disaggregated 6G access networks into a rich service provisioning and access platform. Hyper-local services such as augmented reality scenarios will not require connectivity with a distant service platform. Instead, they will perform better with local real-time time-sensitive service access. Furthermore, individuals will support parts of shared information processing and edge intelligence networks that address collective problems for humanity, such as genome sequencing, through shared resources (à la citizen science). The individual will emerge as a node in the network of intelligence relations, rooted in the local physical world while connected to the hyper-real 6G intelligence networks by an ecology of information devices, products, and IoT services.

Society's need for trustworthiness with associated legal and normative action such as IT security and *privacy* laws are prerequisites to assure the full value and benefit of communications in the 2030s for society and the economy. Privacy regulations need to be strongly linked to the rising trends of the platform data economy, p2p sharing economy, intelligent assistants, connected living in smart cities, transhumanism, and solutions such as digital twins' 'meta' reality. The physical world is increasingly being twinned with the digital world and we are relying more on predictive models to guide collective actions. The 'I own my data' concept is expected to grow, particularly in Europe, based on GDPR evolution. However, severe differences in global data privacy laws are expected to emerge. For example, the US is unique among major countries in lacking a unified set of data privacy laws in spite of having a large number of global webscale companies, and China's cybersecurity law (CSL) applies not only to conventional data handlers but to telecom, radio, and television operators. This is unique because the Chinese authorities must be informed if data indicates any prohibited activity.

Environmental

Sustainable material will contribute to the *innovating to zero* and circular economy megatrends. Toward 2030, companies will shift focus, developing products and technologies that innovate to zero, including zero-waste and zero-emission technologies. The full lifecycle carbon footprint of the ICT industry represents around 2% of worldwide emissions and is projected to grow at a 6% annual compound growth rate. 6G net positive impact and sustainability are expected to be achieved by enabling increased efficiencies and improved environmental performance in other sectors. Computing technologies will be miniaturized to the extent that they will be sustained by the power generated by everyday human activity. Everyday walking, jogging, and housework will produce the energy to support the person's information devices, which will in turn occasionally monitor the person's vitals, as well as cater to information and entertainment needs through over-the-top connectivity.

Key Uncertainties for the 5G/6G Transition

Next, the key uncertainties for 6G were identified based on the ranking of forces according to their high impact and high uncertainty as shown in Fig. 3.2. They are further discussed in the following.

Political, Legal, and Regulatory

AI/ML will provide a new foundation for 6G radio air interface design and optimization, enabling self-optimizing transmitters and receivers, cognitive spectrum use, and context awareness. There are contrary interpretations of the trend in *artificial intelligence rights*. Assuming the availability of appropriate datasets for training purposes, artificial intelligence will be able to propose solutions to increasingly complex problems that can serve as the source of great economic growth, shared prosperity, and the fulfillment of all human rights. 6G will give our networks the ability to sense things. The 6G network is expected to become a source of situational information, gathering signals bouncing off objects to determine their type, shape, relative location, velocity, and perhaps even material properties. It will become capable of determining a person's precise position within a room, as well as tracking and predicting their habits. As 6G biosensing emerges, the network may know our most intimate health details to monitor our medical conditions and medication levels or even warn us of an imminent heart attack or epileptic seizure. In an alternative future, it could drive inequality, inadvertently divide communities, and could even be actively used to deny human rights.

In the connectivity context, *data regulation* relates to net neutrality that rules Internet access providers to treat all traffic equally, irrespective of the sender, receiver, content, service, application, or the device

UNCERTAINTIES

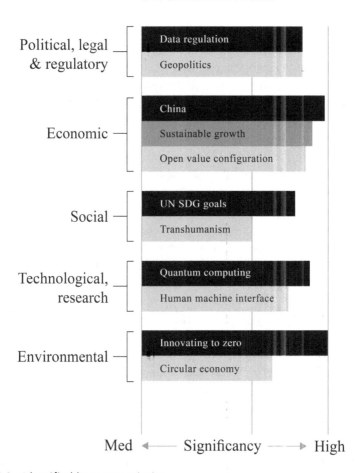

Fig. 3.2 Identified key uncertainties

in use. At the same time, the 5G evolution is already developing a 5G-advanced network that can be extremely tailored to specific use cases intending to treat traffic differently for each use case. This legislation creates uncertainties by impacting companies' capabilities to create and capture value in virtualized network-based services between telecom operators and cloud providers. One of the key uncertainties concerns how edge computing should be provided under strict net neutrality, e.g., as in Europe. Furthermore, it impacts capabilities of providing the cybersecurity required for the merging of the physical and digital worlds that is now happening.

In *geopolitics*, the tension between globalization, networking power, and the urgency of ecological reconstruction will be linked to the balance between centralized decisions and the strengthening of inclusion and democracy. Toward 2030, the power configuration may be transforming from a multi-polarized world to a poly-nodal world in which power will be determined in economic, technological, and cultural networks and interaction. Political and societal systems will face growing tension in responding to the instability of the financial situation, the ecological sustainability crisis, and uncertainty about the complexity of the future world. Societies may struggle to find a balance between fast-moving decision making, community engagement, and the reasserting of democratic values and commitments. On the one hand, it is hoped that strong leaders will bring simplicity to complex problems, but on the other, there could be increasing efforts to influence things in communities from the grassroots level. Furthermore, along with increasing polarization, the aging population in developed countries, and diversification, new tribes and communities will emerge around various imaginary groups representing a wide variety of values, places of residence, political opinions, consumption choices, or lifestyles. It may happen that weakened and fragmented future prospects, the absence of togetherness, and the polarizing effect of social media will lead to a rise in populism, skepticism toward changes in the environment, and in the worst case, extremist attitudes. At the same time, environmental awareness among people and companies may increase and be reflected in a growing number of people and communities changing their habits, and companies taking corresponding actions to offer customer experience. Vehicles of open value creation and the open-source paradigm, in particular, may provide a powerful avenue to reinvent civil society's participatory process in conjunction with anticipatory regulatory capability.

Economic

Resource orchestration and configuration relate to power over development and adoption of innovations and technology that are essential for *sustainable growth* and ubiquitously embedded in society and daily life. Data is increasingly accumulated, and its value and significance are growing. Technology may increasingly be seen as a geopolitical issue of power, and questions of future resource orchestration power will emerge, exemplified by the following three questions. Who will own the continuously accumulating data? Who will get to decide on technology? Who will set the rules and regulations?

Open ecosystems will foster economies of scale, reducing costs and accelerating services. A new landscape of vendors will enable them to become digital service providers. *An open value configuration* will emphasize value co-creation and co-capture to maximize the overall ecosystem value, which in turn may increase the value shared and acquired not only by a focal firm but by the actors within the ecosystem. By utilizing the sharing and circular economy, co-creation partners will employ existing resources and processes to promote stable interaction. Toward 2030, platform ecosystems will not only offer search, social media, and ecommerce but provide an infrastructure upon which innovation and transaction platforms will be built. Platform business models will rapidly overtake and disrupt traditional linear value chain business models. 'API-fication' as the confluence of a few different trends such as ubiquitous software, cloud computing, and microservices will enable new digital value flows in all industries, e.g., marketplaces to lease/share resources and to make the capabilities of personal subnetworks and enterprise private networks available.

Digitalization and software will be everywhere, and building a developer ecosystem will be a pre-requisite for success for many businesses. Novel decentralized business models will not necessitate a focal point but will require the design of transaction content, structure, and governance to create value. Everything-as-a-service will become the dominant model beyond IT and will evolve to outcome-as-a-service providing a service-level agreement (SLA) and will be on-demand with elastic access to applications, information, and resources.

China was seen with high uncertainty in all categories. On the one hand, China is becoming a leading superpower, exploiting foreign resources and technologies, while the Belt and Road initiative offloads excess capacity of produced commodities. On the other hand, it is threatened by its aging population, pollution, and political instability which may turn China inwards to maintain peace and prosperity and bring its economic progress to a standstill. Megacity pollution in China will drive green tech forward. China is investing heavily on technology leadership aiming to set the standards and the technical direction for the entire world. Chinese politics and technology will become strongly intertwined as technology such as new IP systems support the Chinese political structure. This government led, enterprise-driven approach will create unique super-platforms based on pure commercialism and will be implemented with a mobile first mindset. The Chinese market will differentiate itself strongly from the global market in terms of its offerings, services, use of technology, and market regulation.

Social

Commitment to *sustainability* principles will be confirmed with clear targets and performance reporting, e.g., UN 2030 agenda for sustainable development; corporate responsibility; ecological, social, and governance. Novel sustainability-oriented business model innovation will emerge stemming from the social freemium, circular economy, product-service, and sharing platform business models. Both the commercial launch of 6G and the United Nations' sustainable development goals (UN SDGs) are targeted for 2030. 6G communications are expected to boost global growth and productivity, create new business models, and transform many aspects of society. The UN SDGs are a way of framing opportunities and challenges of a desirable future world, and they cover topics as broad as ending poverty, building gender equality, the fight against climate change, and developing smart cities.

Transhumanism reflects the rise of technology-driven evolution at an unprecedented rate of change, prompting deeper questions into what it is to be human from the biological, behavioral, and human–machine evolutionary perspectives. By 2030, we could see a greater societal focus on sustainability, the nature of humanity, values, creativity, and self/social fulfillment and empowerment (Kinnula & Iivari, 2019). There may also be a cognitive intelligence revolution via the ascendancy of sentient tools and possibly also a human-orchestrated self-directed selection in biological, neurological, and physical evolution. Human augmentation will evolve from primarily gaming applications to everyday value-added mixed reality (XR) services offering safety, productivity, and efficiency improvements from hybrid working worlds to industrial processes. Furthermore, the metaverse will extend virtual digital twin models from technologies allowing an interactive experience toward digital avatars. The 6G metaverse will enable us to interact with each other, create, and utilize virtual resources and services, and experience a variety of new features.

Technological and Research

Alternative computing approaches such as *quantum computing* will shine at sorting, finding prime numbers, simulating molecules, and optimization and could thus disrupt segments like finance, intelligence, drug design and discovery, utilities, polymer design, AI and big data search, and digital manufacturing. The technology may for a long time be limited to selected industries such as the military, national laboratories, and aerospace agencies, while alternative computing approaches to help handle the greatly increasing level of parallelism in algorithms may be available more widely. Today's asymmetric cryptographic algorithms will likely need to be replaced with quantum-safe concepts to provide a new approach to secure 6G networks and protocols.

Environmental

The circular economic model represents a shift from constantly producing new goods to a form of consumption that is based on using services and sharing, lending, and recycling goods, instead of owning them, aiming to reduce waste. Digitalization offers many opportunities for monitoring and steering the circular economy and understanding the big picture of the sustainable data economy. The constant progress of technology, 6G, computational capacity, cloud-based services, and Industry 5.0 will make it possible to make many factors of production 'intelligent'. This, in turn, will improve efficiency, monitoring, and make processes more transparent and facilitate the development of new kinds of personalized digital services. 6G aims to create zero-carbon-footprint networks where every aspect of the network's operation is designed to minimize or offset CO_2 emissions. Not only will 6G enable lower costs per bit and faster connectivity, but it will also be able to analyze collected data to turn off components and scale down capacity when the demand is low. Energy efficiency will be a major design criterion in 6G along with the other metrics.

6G Impact Scenarios Based on Key Trends and Uncertainites

Based on the collected data and identified trends and uncertainties, the following eight scenario logic dimensions, referred to as themes, and their endpoints were selected to develop future 6G business scenarios (Yrjölä et al., 2020), as summarized in Fig. 3.3. The identification of dimensions was done by choosing key trends and uncertainties based on their anticipated impact and the predictability of consequences and uncertainty. The scenario logic was selected to represent the most significant uncertainties of the overall system under scrutiny by selecting two unrelated polar dimensions. The eight scenario dimensions were categorized into four themes to develop a total of 16 alternative 6G scenarios, which are discussed next.

In the following sections, the scenarios are presented top down from the user and developer level to firm, ecosystem, and geopolitics level as depicted in Fig. 3.3. The description and numbering of the scenarios are given in the scenario titles and shown in Fig. 3.4.

SCENARIO THEMES

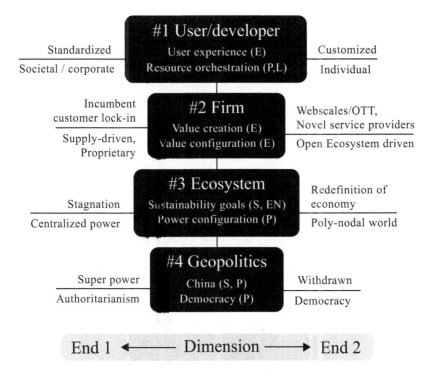

Fig. 3.3 Summary of selected scenario themes

User and Developer Level Scenarios

In the user developer scenario theme, the horizontal dimension was chosen to represent resource orchestration and the vertical axis shows the user experience. The polar dimensions on the resource orchestration axis range from societal/corporate to individual-driven orchestration. The user experience axis ranges from traditional standardized service provisioning to the opposite driving customer engagement with customized long tail service experiences. Using these two scenario dimensions, we have developed four scenarios.

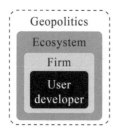

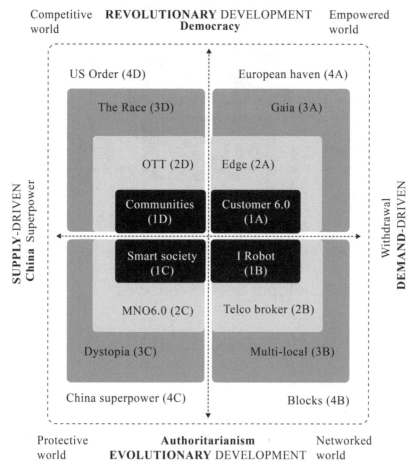

Fig. 3.4 Summary of four scenarios

Customer6.0: Customized Experience and Resource Orchestration by User (1A)

In the Customer6.0 scenario, the user experience is customized, and resource orchestration is user and developer centric. In this scenario, 6G technology has penetrated most parts of the world. IoT devices and sensors controlled by AI are a normal part of the environment nearly everywhere. The automatic collection of different kinds of data from humans, as well as from our environment and its analysis, is used for highly sophisticated products and systems that make people's lives easier and provide a better user experience through convenience, because everything is automated. The prices of the systems are very reasonable due to open interfaces and standardized cheap components. Computing technologies are miniaturized to the extent that they are sustained by the power generated through everyday human activity to support the individual's information devices, which in turn occasionally monitor the person's vitals, as well as catering to information and entertainment needs through over-the-top connectivity. Media and service consumption are rooted in local economies, and users of such products and systems are used to living with them and cannot imagine their lives without them.

Furthermore, in this scenario, hyperconnected and completely automated networks have been put together with a public–private-personal ownership funding model with a view to sustainable growth and digital infrastructure usage. As counterforces to the creation of platform monopolies, decentralized platform cooperatives, the peer-to-peer economy, shared economy models, and the progress of a human-driven fair data economy have emerged. Transhumanism reflects the rise of technology-driven evolution at an unprecedented rate of change, prompting deeper questions about what it is to be human from their biological, behavioral, and human–machine evolutionary perspectives. By 2030, we could see a greater societal focus on sustainability, the nature of humanity, values, creativity, self-/social fulfillment, and empowerment. There may be a cognitive intelligence revolution via the ascendancy of sentient personal assistants and possibly human-orchestrated self-directed selection in biological, neurological, and physical evolution. The emerging opportunity to redefine human–machine and brain-UI interfaces enables the connection of people and the biological world in novel ways. Holo-presence systems can project realistic, full-motion, real-time 3D digital twin images of distant people and objects into a room, along with real-time audio communication, with a level of reality rivaling physical presence. Images of remote people and surrounding objects are captured and transmitted over a 6G network and projected using laser beams in real time.

There will be a serious threat of a digital divide and inequality related to access and skills to use new technologies, knowledge, digital services, and materials at the individual level, as well as between countries, if access to new technologies is restricted. This may reflect working careers, which also assumes employee activity to educate individuals. Expensive products and systems may never be the norm, but both local businesses and citizens can create their own frugal adaptations of products and systems to suit their living conditions when even electricity may be a scarce resource. This is supported by the global developer culture, in which the sharing of blueprints and working processes is encouraged. Development talent will become a scarce resource, and most companies will be struggling to attract the talent they need. The shift to cloud-based services has changed how enterprises purchase software and its development. Application developers will have more control than before over what is being purchased. Companies will build their products to make it easy for developers to adapt and shift their expensive top-down go-to-market motion to bottom-up product-led growth, where customers can easily try out the product and expand usage over time. Open supply of best-in-breed (SW, HW, services) in a decomposed and open architectural environment with open interfaces and open hardware is adopted.

6G will transform urban and rural living at the intersection of geopol-itics, the growth of nationalism, rights to information transparency, and information citizenry. Content growth will lie in supporting people's multiple social and technological identities through a variety of media. This will require a mindful view of decision making and the regula-tion of future data, information, media, and network usage in light of sustainable growth for the economy and human empowerment.

I Robot: Standardized Experience & Resource Orchestration by User (1B)

In the I Robot scenario, the user experience is standardized and resource orchestration is user centric. With the convergence of nanotechnology, biotechnology, information technologies, and cognitive science (NBIC), newer application areas, goods, services, and systems will proliferate. This convergence has resulted in our present development of cyber-physical systems and IoT-based technologies, along with 3D printing and on-demand manufacturing, among other instantiations. In the next two decades, the growth of 6G-enabled technologies will aid the explosion of biologically based intelligence along with artificial intelligence in industrial setups. This biological intelligence will rely on a mixture of biologically based self-programmable natural and artificial neural networks and micro- and nanobots that can be used in tandem with existing AI-based automated systems. Interaction with these hybrid bio-industrial automated systems will constitute the next major revolution in programmable smart factories and industrial systems, and it will be a source of value creation, configuration, and resource utilization. In a rapidly changing reprogrammable and reconfigurable world, businesses will have a short to medium change horizon, expected performance indicators, and return on investments. The market of mid-level businesses in the industrial sector will be increasingly facilitated via agile and scalable techno-parks and spaces. The people, processes, and resources needed for such mid-scale businesses and services will require flexibility and rapid learning to transfer learning from one job to another.

While countries will continue the movement of goods in a globalized world, the nature of transfer will have a marked impact. In a world enabled by 6G technologies and 3D printing, the blueprint will be delivered to proprietary machines, which will print products as required in a model of edge-based manufacturing, with designs supplied and monitored through remote setups. Local-demand–supply–consumption models will become prominent in an already globalized world, with a marked emphasis on localized spatial circular economies. To ensure independence, assurance, and resilience, local manufacturing will be decentralized to several manufacturers, which will together compose a crowdsourcing production ecosystem. Managing the ecosystemic network of small manufacturers will utilize blockchain technology for supply chain management, smart contracting, and transactions. It will be possible to move production sites to new locations, enabling remotely controlled worksites and heavy-duty vehicles. Production models will be driven by sustainability, resilience, and the circular economy.

I5.0 will allow collaborative human–machine interaction with robotization across services and industries, because human intelligence will be in perfect harmony with advanced cognitive computing. With real-time data, effective data monetization, and digital automation of the manufacturing process, businesses will be able to shift their focus to generating higher revenues from the servitization of products. Open interfaces and advanced manufacturing capabilities will help to overcome design complexities. With its ability to facilitate an extremely long tail of mass customization, it will also return control to customers in a haptic way. Furthermore, trustworthy quantum-enhanced I5.0 networking and services will provide the highest standards of safety and environmental protection.

The use of programmed organisms will become increasingly common in production. Genetic engineering and synthetic biology will enable the creation of new kinds of organisms, as well as the modification of existing organisms for specific purposes in food production, chemical processes, textiles, and in the pharmaceutical and construction industries, for example. This will decouple growth from cost and resource usage. In interaction with hybrid bio-nano-artificial intelligence, industrial technology operators and maintainers will be forced to adopt a special synchronicity with these technologies, which in turn will be adaptive to the workers. Unlike the industrial revolution of the 1900s in which humans were subjugated to the rhythm of the machine and which has prompted a backlash to the mechanistic life brought about by modern times, the new industry 5.0 technological revolution enabled by 6G technologies will bring about a new rhythm that links the biological dimension of the machine to that of the human. This will impact the human at sub-awareness levels, bringing about heightened nervousness, anxiety, and general discontent.

Smart Society: Standardized Experience & Resource Orchestration by Society/corporations (1C)

In the smart society scenario, the user experience is standardized, and resource orchestration is society/corporation centric. Technology will develop rapidly, changing production methods and operating models. A growing number of things will be automated, production and operations will be decentralized, and interaction will often take place remotely or via a virtual environment. This assumes continuous learning from individuals to keep track of development and evolve their professional skills. Making use of technology will increasingly call for changes in thinking models and operating methods. The gamification of working life may offer motivation for some people frustrated by the changes. Smart society will build dependable systems and communication in which standardized data is utilized by walled garden platform monopolies across verticals. The smart city focus will be extended to rural inclusion. Multi-locality will be the norm in combining city life and isolation from crowds. 6G will transform urban and rural living at the intersection of geopolitics, the growth of nationalism, rights to information transparency, and information citizenry.

A dependable communication system that allows remote work and telepresence in real-time mode will result in knowledge-based jobs and other net-based service sectors shifting to a bucolic life in which urban and rural life remains in healthy balance. The change in lifestyles will enable an emphasis on collaboration for the common good and making society more inclusive of the requirements of disparate cultures and sub-cultures. In this regard, there will be a marked shift to appropriate data and privacy regulation to support vested interests and motivations. In this smart economy, consumer insights, virtual finance, carbon-free consumption, low energy consumption, and global and fiscal sustainability will take center stage. Thus, there will be an expansion of the social intangible economy, involving several types of online gaming, social media exchange, interaction in virtual holopresence interactions, and other forms of digital currency exchange. There will also be a rapid convergence of these various interactions, such as making online groups, communities, and institutional rules that will assist in creating an information citizenry and a reciprocal impact on real-world global issues.

The most important global concern will be to ensure mutual respect for people from every stratum of society. This will be possible through digital inclusion in all sectors, ranging from finance to education. The aim will be to create a just and egalitarian society using appropriate information regulation and mutual distancing through the creation of safe and creative collaboration spaces that support the interests of like-minded groups. Actions at the level of individuals supported by 6G technologies will provide a morally sustainable world in which every citizen will be a self-aware informed citizen with a dual identity: recognizing allegiance to the nation, as well as living within the constraints of the global pan-dimensional virtually connected world.

Communities: Customized Experience & Resource Orchestration by Society/corporations (1D)

In the communities scenario, the user experience is customized, and resource orchestration is society/corporation centric. The sense of community created by 6G technology and the ability to directly collaborate with others will enable humans to participate and act in society in an unprecedented way in countries where access to new technologies is the norm due to competences and skills in using new technologies. The sharing economy, crowdsourcing, and crowdfunding will expand the space for new forms of organization and innovation.

In this heterogeneous society, social networks and the trust and reciprocity they foster will be highlighted from the perspectives of well-being and working life. Public network funding has traditionally been directed at unserved and underserved areas in terms of broadband access and coverage. Support for deployment programs will be extended to policy areas such as smart city community development, worksites, and ecosystems (such as harbors and airports), advanced health services, logistics and transport, public safety, and critical infrastructure at length. Hyperconnected and completely automated smart grids will be extended to a variety of vertical sectors, including electricity, the Internet, and health care, serving as a middle layer between humans and natural environments and enhancing the capabilities of both. These networks will have been assembled with a public–private partnership funding model, with a view to sustainable growth and digital infrastructure usage. The human body will be a vital part of the Internet of Senses. Increased data will enable more personalized and preventive care in which AI-assisted analyses will monitor personal indicators and compare them with larger population data, offering medical consultation via a specific indicator trigger point. Biological processes and communication systems will be integrated with technical communication systems, providing online information about vital transactions and guiding us to take specific actions to remain healthy. In the event of infection, we will receive continuously updating diagnoses to be shared in real time with healthcare professionals, who will base their consultancy on AI-driven analyses. New treatments will also be developed based on genome editing and modifying the microbiome, for example.

Countries with less restrictive legislation will act as resource pools for corporations by providing cheap labor forces, natural resources, and (private) data about humans (use data, biodata, biological data, etc.). Frugal innovations will be developed to serve the growing customer base in low-income countries. Education powered by the Internet of Skills and the Internet of Senses will enable specialization from the school system's early grades. Learning will be tightly connected to personal data to react to any disturbance and ensure a successful study track based on individual interests. Students will be able to choose virtual courses and degrees from any university globally and visit digital twin campuses for interaction. Global networking during studies will support international career planning, which is done partly remotely.

Several ethnic communities will have struggled to maintain their existence in everyday and virtual spaces. The nature of communities will change in a variety of ways and on varying timescales. This will mean:

the fragmentation of communities; dynamic tension between individuals and communities; the morphing of community values and identity; and other phenomena. Radicalized groups will have emerged, spreading terror both online and offline. The spread of cyberterrorism may affect every networked system in the world, resulting in a global crisis and a devastating effect on the world economy. In the wake of disasters (terrorist attacks, tsunamis, diseases, etc.), 6G technologies may also support the victims. The growth of human-body-powered networked devices will help communities to establish informational relations that aid the troubled, enabling the community to show resilient behavior and bounce back quickly.

At the level of communities, media interaction will result in the intensification of activities related to public opinion shaping. These will include the transmission of hate speech and fake news, which will also be experienced somatosensorily. This holistic experience of various forms of malevolence will have a much stronger impact than ever in mobilizing people toward crime and terrorism though virtual technologies. Special interest online communities will continue to proliferate. However, with 6G experiential technologies, these special online communities will move toward a more accelerated and hyper-real set of interactions. Hate speech and associated activities will not only be symbolic but tangible. A final twist in the life of communities propelled by 6G technologies will be in terms of the 'wisdom of crowds'. In normal circumstances, this 'wisdom' will allow for more egalitarian and informed decision making and empowerment. However, with the hyper-real experiential hate of 6G-enabled vitality, 'wisdom' may be perverted without bounds, resulting in a bleak communal life.

Firm-Level Scenarios

In the firm scenario theme, we chose the horizontal dimension to present value configuration and vertical axes value to capture logic, as depicted in Fig. 3.4. The polar dimensions to the value configuration axes are a traditional closed supply value chain focus and open ecosystemic-driven configuration. Value creation customer attraction and lock-in logic form the incumbent mobile operator-dominated model; the opposite is the expanded model with webscale companies, OTT, cloud, I5.0, and novel digital service provider stakeholders.

Edge: Value Creation by Novel Service Providers & Open Ecosystem Value Configurations (2A)

In the edge scenario, value creation is customer attraction- and lock-in-driven, and value configuration is open ecosystem focused. This scenario stems from a decentralized open value configuration and ecosystem-driven business models. Modularity enables disaggregation and decomposition. Collaborative standardization will still be needed to harmonize and maintain replicability for interoperability reasons. Government funding and public–private partnership (PPP) will influence the telco industry to open and transform next-generation communications as a general-purpose technology. A decentralized platform will distribute the value between the players, while open-source software will lower market entry barriers for developers, promote interoperability, and expedite development cycles based on shared knowledge. Unlicensed common spectrum and novel shared and locally licensed spectrum will enable openness, democratizing markets, and allowing subnetworks in mobile communications. At the same time, value capture opportunity diminishes. It will become harder to profit from technology innovation due to the spill-over effects when moving toward general-purpose technology and further reduced royalty-based revenues from standard or essential patents. Only the best, fastest, and most efficient producers will win (as in the case of webscalers), and companies will need differentiation over and above open source.

Novel players will take over both customer ownership and networks. OTTs and device vendors will own the B2C customer relationship, while local tailored private cloud-native networks will attract B2B customers. Telecommunication operators will play a role as wholesale connectivity service providers. Technology and innovation ownership will be expanded, fully leveraging the open API world and novel resource brokerage. Furthermore, edge resources may be operated by local communities and special interest groups, e.g., by expanding services into remote rural areas, or universities and research organizations deploying their own edge resources to speed up local innovation. Nomadic subnetwork edge elements will provide sustainable and ecologically efficient deployments. For example, communities that do not wish to have invasive technologies in their midst may hire edge-enhanced systems for occasional high-quality capture and streaming of local events. Banks, healthcare centers, and governance points may extend the regular telecom networks for the affordable inclusion of the masses. Onboarding, as well as the billing of customers via digital cash and keys, may be done by these local entities, who in turn may pay the wholesale service providers for wider connectivity. These semi-autonomous 6G subnetwork deployments will be heterogeneous in nature, often encouraging innovative products and services. There will be specific network areas and zones. We may have a personal zone with in-body communication applications producing data for daily diagnosis for individuals or shared merely with a private medical doctor. The wider zone could be a common family network, shared strictly between family members at home and offering tailored services. Moreover, particular network zones will be shared with various interest groups enabling dynamic management based on personal preferences and the changing requirements of groups. Security, trust, and identity management in such heterogeneous edge deployments will be both a challenge and opportunity for novel, sustainable business models.

Telco Broker: Value Creation by Incumbents & Open Ecosystem Value Configurations (2B)

In the telco broker scenario, value creation is driven by incumbents—the existing operators—and value configuration is open ecosystem focused. Telco brokers will have retained the primary customer relationship and will have focused on monetizing data via the service platform ecosystem. Technology providers will drive the technology ecosystem and offer an efficient network infrastructure via platform-based ecosystemic business models. The decoupling of technology platforms will have lowered the entry barrier, allowing multiple developers and entities to contribute to the innovations envisaged with 6G. Moreover, fine-grained modularity and open source will allow highly specialized long tail solutions and services from smaller payers to be widely deployed, leading to innovation, and possibly to commoditization.

With open interfaces, software functionality will become modular, allowing developers to effortlessly continue to develop a solution by adding, changing, or removing components with much more speed, agility, and productivity than traditional, monolithic applications have provided. Consequently, the theory of the firm, meaning the nature and structure of a company, will need to be thought over, including how companies are organized internally and where the boundaries are between the company and the market. API-fication will break scale economies of end-to-end platforms with best-of-breed solutions. APIs will change how software is developed by allowing companies to mix internal and third-party components when creating an offering. This will be the key driver for platform interplay and decentralization. Distributed ledger technologies will develop rapidly and enable decentralized data storage, computing, and AI. As Web 2.0 created cloud computing and today's platform businesses, such as Meta, Google, and Amazon, Web 3.0 is all about edge computing and decentralization. It will fundamentally change the characteristics of networks, ownership structures of companies, and incentive models. It has the potential to bring back the benefits of the early Internet, such as openness and peer-to-peer (users having the power vs. large companies), while still being scalable and efficient. Networks will become open, trustless, permissionless, interconnected, secure, and monopoly resistant. Decentralized autonomous organizations (DAOs) may be beyond regulatory reach while changing business cases for webscalers.

MNO6.0: Value Creation by Incumbents & Closed Ecosystem Value Configurations (2C)

In the MNO6.0 scenario, value creation is incumbent-driven, and value configuration is closed supply focused. In this scenario, telecommunication firms will drive technology innovation and control the traditional e2e value chain, owning the customer relationship in both B2C and B2B segments. It will be strongly under business-driven decisions by MNOs that the advanced services enabled by 6G technology will be available for various verticals. Key technology enablers to be utilized as the prerequisites for the decoupling of costs from growth will include automated network slicing and leverage, and the use of higher frequency bands in conjunction with network densification. In addition to the technology innovation platform, there will need to be a transaction platform position between customers and OTT players. This tightly coupled deployment may provide optimal efficiency with respect to efficiency, environmental sustainability, and technology exploitation in the connectivity layer. Via the opening up of network interfaces, telecom firms will co-develop within their value chain and will use open-source software to address the long tail of specialized local and industrial use cases.

The business environment will become unpredictable with high geopolitical uncertainty, increasing inequality, and climate change. With the mainstream managerial practice focused on managing performance, resilience will represent not just an opportunity to mitigate risks but also an opportunity for competitive advantage. Disaster recovery of critical infrastructure will become a growing business segment requiring co-creation between industries to overcome the financial constraints. Market opportunities will exist particularly in secured networks such as security and privacy as a service and blockchain networks. The need for modularity to reduce the dependency on single vendors may reverse the vendor consolidation trend seen in the past and lead to more fragmentation. With emerging safe quantum technologies, security solutions and encryption will become geopolitical control points.

Exponentially reduced costs related to satellite launches and production will enable cash-rich companies to fund LEO constellations (e.g.,

SpaceX Starlink, Amazon Kuiper project). The impact on mobile broadband will depend on the ability to integrate this with the terrestrial cellular terminal ecosystem. Sustainability value may be created via building a fully solar-powered communication network with virtually 'zero' energy consumption and innovations enabled by sensing via satellite. Global satellite constellations will represent geopolitical control points. Effective policy and regulations for orbital consideration will be needed to avoid potential collisions in space and to limit the impact on astronomy, and spectrum allocations will be required to ensure harmony between terrestrial and 5G networks.

Over-The-Top: Value Creation *by* Novel Service Providers with Closed Ecosystem Value Configurations (2D)

In the over-the-top scenario, value creation will be customer attraction- and lock-in-driven, and the value configuration will be closed and supply focused. In this scenario, webscalers and OTTs will have taken over customers from telecom operators by utilizing their access to customer data. However, operators will continue to control both the mobile and fixed connectivity technologies that will be commoditized and will orchestrate the related end-to-end (e2e) value chain. Commoditized connectivity will drive operators to create partnerships with webscalers, OTTs, I5.0 service providers, and public networks and to provide wholesale services utilizing their cloud and transaction platforms. OTT players will offer novel free or subsidized connectivity business models, utilizing revisited net-neutrality principles, affordably expanding their reach to the bottom four billion.

MNOs will lack a global developer ecosystem, application-level knowledge, and AI capabilities and will see webscalers as a lever to monetize IoT and edge technology. The risk is that their role will be reduced to that of a last-mile connectivity provider or 'just another vertical' but with needed assets such as spectrum and edge real estate. Webscalers will drive the value chain in the edge application context and even create a new revenue source with hyper-local cloud infrastructure service with scalability, required availability, and almost unlimited flexibility. The diminishing MNO value share in edge cloud deals will trigger private wireless deals by webscalers bypassing MNOs and leveraging the MNOs' infra-assets and creating a service layer to limit their value capture. Webscale companies will drive their successful transactional platform business model into new/adjacent domains where winning platforms will cover innovation and transaction. Companies with a traditional linear business model will become less attractive to investors. Payment models will shift from CAPEX to OPEX and business models from transactional to continuously digital-connected, creating new collaborative models and interdependencies. Pressure on cost and risk mitigation will increase while technology complexity will continuously grow—with lower entry barriers for newcomers and increased competition for incumbents with a tendency to become a winner-takes-it-all market. The outcome-as-a-service trend implies that service providers will share liabilities, as well as the need for information security, data privacy, and adherence to compliance and regulations. Resilient and secure network connectivity will be based on clearly defined service-level agreements.

There will be continuous pressure on MNOs to generate higher returns: government regulations may separate infrastructure firms and operation/service providers; network sharing deployments will optimize site needs and reduce the market initially; and finally, no government regulations will prohibit webscalers and neutral hosts from becoming service providers. Furthermore, the edge cloud will trigger a second round of investment, triggering neutral host and webscale alliances that will transform MNOs from telcos to tech comms companies. Neutral hosts will transform from cost-optimizing units to new wholesale business units.

Ecosystem-Level Scenarios

The sustainability crisis, which refers to the deterioration of the environment and exceeding the earth's carrying capacity, may significantly change our operating environment as we move into the 6G era. The ecosystem scenario theme particularly recognizes the UN's SDGs as important drivers for 6G, but the approaches vary. Responding to increased environmental awareness requires changes in culture and practices and has been accompanied by a polarization of views. Hybrid military, economic, technological, and cultural powers have become overlapping, leading to threats and hybrid influence. In the sustainability scenario theme, the horizontal dimension represents the power configuration. The vertical axis represents sustainability development. The polar dimensions opposite to the power configuration axis are centralized power and poly-nodal configurations. The sustainability dimension ranges from the redefinition of the economy and its opposite, stagnation.

Gaia: Sustainability by Redefinition of the Economy & Poly-Nodal Power Configurations (3A)

In the Gaia scenario, sustainability is driven by the redefinition of the economy, and power configuration is poly-nodal and world focused. Environmental awareness among people will have increased and resulted in corresponding actions. Dissatisfaction with the current measures taken with respect to climate change and biodiversity will have motivated a growing number of people to voice their opinions and participate in demonstrations. Instead of individual poles of power, the emphasis in global politics will be on relationships and interaction. In addition to governments, other players, such as businesses, lobbyists, think tanks, international institutions, cities, and activist organizations, will play a significant role in this. In the 6G-enabled real-time economy, all the transactions between business entities will be in digital format, generated automatically, and completed as they occur without the need for storage and forward processing. In innovating to zero, the cost of renewable energy and storage will fall. Energy production will also become increasingly decentralized as more and more people produce their own energy and sell what they do not need. In the circular economy, production and consumption will be planned to prevent waste from being generated, while materials and their value remain in circulation via sharing, leasing, repair, and reuse. In this context, deploying 6G could entail an application area to provide ultra-low-power communications through energy harvesting or wireless power to very small devices, for example. The counterforces to winner-takes-all monopolies will include platform cooperatives, the peer-to-peer economy and sharing economic models, and the progress of the human-driven fair data economy and the fair distribution of wealth. This restorative economy will lead to a society characterized by broad empowerment, greater equality, a higher level of well-being, and better sustainability. The Internet of Senses and the Internet of Skills will utilize advanced human–machine interfaces to enhance the human intellect and physiology toward transhumanism. In the Gaia scenario, societal resilience will provide the ability to cope with and overcome adversities, the ability to learn from past experiences and adjust to future challenges, and the ability to craft sets of institutions that foster individual welfare and sustainable societal robustness to future crises.

Multi-Local: Sustainability by Stagnation & Poly-Nodal Power Configurations (3B)

In the multi-local scenario, sustainability is stagnation driven, and the power configuration is poly-nodal world focused. A shift will have happened in global politics moving from a multipolar to a poly-nodal world. The geopolitical power blocs—the US, Europe, and China—will give way to a networked world with nodes comprised of countries, emerging economies, corporations, and other non-state actors. In the face of disruption, people will turn to increasingly polarized tribes and bubbles formed around values, place of residence, political opinions, consumption choices, or lifestyles for guidance. The fragmentation of the economy, transformation of work, and new organizational models of the sharing and platform economy will have challenged the traditional relationship between the employer and employee, and what the benefits are. Working life will become increasingly diverse, and there will be an emphasis on ensuring people's livelihood and competence building. Universities will offer tailored virtually augmented education environments to enterprises and private entrepreneurs. Consumers will favor domestic or local products and can themselves choose where and how goods and products are manufactured. 3D printing will allow many products to be manufactured at home or in the neighborhood. Distributed local production will on the one hand be practiced by enterprises using a network of geographically dispersed manufacturing facilities coordinated using 6G, while on the other hand, it will also be practiced by local developers and manufacturing. Traditionalism will develop as a response to disorder and will favor public–private partnerships.

Dystopia: Sustainability by Stagnation & Centralized Power Configurations (3C)

In the dystopia scenario, sustainability is stagnation driven, and the power configuration is centralized and power focused. We will continue to inhabit a consumption culture in which nature is seen as a free resource that we use as we wish. The wealth generated by economic growth will not be distributed sustainably and will be concentrated in the hands of a shrinking minority. Occasional large disasters will not make wealthy people act if they are not threatened themselves. Weakened future prospects, the fragmentation of the political map, and the polarizing effect of social media will have led to a rise in populism, which emphasizes the division between the elites and the masses. The benefits of internationalism will not be acknowledged, because they are considered too indirect, and its negative aspects will be emphasized in the discussion. Globalization will also have led to an opposing reaction in the form of increased nationalism and an emphasis on national borders and will favor state corporations. Democracy will be challenged by ideas of practical autocracy and technocracy, as well as by the notion that democracy is too slow or ineffective to respond to the urgent questions of our time. The need for rapid major changes and a yearning for simple solutions will make strong leaders more popular, presenting a challenge to individual freedom and democracy. The amount of disinformation will grow, and efforts to influence opinions will be increasingly geared toward instigating confusion and discord. A digitalization backlash will have happened, and people will rage against machines.

The Race: Sustainability by Redefinition of the Economy & Centralized Power Configurations (3D)

In the race scenario, sustainability is redefined as economy-driven, and the power configuration is centralized and power focused. The urgency of climate and sustainability action will have led to eco-dictatorship and creative destruction because a voluntary change in people's behavior will be considered so unlikely. The population will have become concentrated in a small number of growth centers, where vibrant megacities and unicorn superstars dominate innovation addressing individual technologies and the ecosystems they form. The process of innovation competition will incessantly revolutionize the economic structure from within, giving time for market advantage.

Geopolitics-Level Scenarios

In the fourth theme, the geopolitics theme, the scenarios discuss how political power is reinforced or undermined by geographical arrangements like boundaries, coalitions, spatial networks, natural resources, and technologies. Furthermore, the current era of geopolitical overrides the role of democracy in the international order. Recent years have witnessed regional and global power plays by Russia and China. Their international efforts are usually cast as moves to establish spheres of influence, but they are broader than that. In the created scenarios, the vertical axis represents democracy, and the horizontal axis represents the influence of China, as depicted in Fig. 3.4. The polar dimensions of the democracy axis are democracy and authoritarianism, while the axis describing the role of China ranges from superpower to withdrawal.

European Haven and the World of Blocks (4A,4B)

Slowing globalization on the one hand and the demands for economic resilience on the other will have led to the formation of three distinct blocks: Europe, North America and East Asia. The blocks will be largely self-contained economies which trade with neighboring areas, but not very much with the other blocks. How each block approaches various issues reflects its cultural traits. The US, a horizontal society, will clearly be about the individual and people's own choices and will focus heavily on the consumer. China could be best described as a three-dimensional actor, meaning that the state will be involved in everything. Europe, on the other hand, will generally be vertical; organizations will play a key role in both planning and development.

China will have been moving some of its manufacturing base abroad, including to Africa. Largely self-reliant in innovation and product development, the country will now be facing the way of Japan; the aging population will threaten to bring its economic triumph to a standstill. Financial woes will start to set in as the economic growth, albeit still higher than in the Western countries, cannot keep up with public spending. The US will be mired in its internal problems, and its debt-laden economy will be a subject of constant predictions of imminent and disastrous collapse. Western Europe, on the other hand, will also be struggling to maintain its welfare systems.

By 2030, EU will still be known for many positive things, but its main competitive edge will be its reputation for its transparent and uncorrupted government. European countries will have been leaders in sustainability and inclusion. They will now be concerned with the security of supply chains, energy, food, and defense. The competition in the field of technology will now mainly take place between the US, EU, and the joint efforts of Japan and Korea. China will still create significant numbers of innovations, but the totalitarian society will not to be able to grant the space needed for truly unique developer-led innovations. The situation in China will be further exacerbated by worsening corruption made possible by the failed introduction of AI-powered governance, an initiative that remains mostly a pet project of some elite members but with little real-life implications. Europe will have emerged as a haven of both individual rights in the online world and a hub for open-source endeavors. The EU will have managed to create regulations that have enabled open-source producers to certify their work as reliable. The Union will also have banned some foreign tech giants from its markets because of data protection and espionage issues. Europe will have become the main driver toward a more human-centered society in both techno-logical and governmental questions. Whereas the US will be all about consumer needs and sellable products, while Europe, in addition to the consumer focus, will seek to ensure that the products are ethical, that is, sustainable, and responsibly and fairly produced. Through its research efforts, the EU will have contributed toward a more open society, as, paradoxically, it will have made communications more secure. People will feel that their secrets will not be divulged without their consent, and they will feel comfortable to be more open about those things they are ready to share. The Western governments, as taught by the pandemic, will have been in the process of adopting the principle of openness.

The US and China will be locked in competition and will have some-what eschewed the notion of sustainability to ensure primacy, although lip service to the environment will regularly be paid in public speeches. Europe will also be virtually the only place where the issues related to AI ethics are taken seriously, while the two other blocks will be competing for supremacy in that field without any concerns.

The aging population in Europe, together with the memory of the pandemic and need for cost cuts, will have made telecare a perma-nent fixture of the healthcare systems. Additionally, work as well as entertainment will be taking increasingly place on online platforms, not necessarily anymore in offices, theaters or cinemas. However, this will also create problems with loneliness and isolation. Thanks to the intro-duction of 6G, each block will have created its own system of watching the movements of its citizens to, at least officially, be prepared for the future pandemics. The European car industry will have become a central technology advancing power. Autonomous vehicles will have entered the streets, in both the public and private transport markets. Due to the automotive industry's role, industrial data will become especially valuable in the EU, in contrast to the customer data in the US. Addi-tionally, Europe will be more concentrated with business-to-business products, whereas the US will be driven by the business-to-customer focus. Automation will have taken leaps forward in Europe, since indus-trial work, that will have to some extent returned to the continent, will be expensive.

Each of the blocks will start to diverge in how different technologies are mixed. Europe will focus on the automation of movement and logistics because of the industrial interests of different member states, while China and the US will adopt local production concepts, mainly thanks to their space-faring efforts. The technologies, however, in many cases will not become diffused between the blocks because of protective laws. As the EU will have also been the leading champion of sustainability, regulations enforced by it will now necessitate that sustainability is one of the critical factors in virtually all development plans and public tenders.

China the Superpower: China Becomes Dominant in the Global Economy (4C)

In this scenario, by 2030 the old regime of the post-Second World War order will have been buried for good and China will be the leading economy and becoming the leading superpower. The Chinese belt and road initiative (BRI) will have tied much of Eurasia to China, ousting US influence in many places. Active trade will take place throughout Eurasia, but 'slowbalization' will have somewhat affected the commerce between North America and Eurasia. Because of the stress that the rise of China will have caused, the US will be more and more openly advancing only its own interests. China will be actively seeking to buy and utilize foreign companies with advanced technologies. At the same time, it will still send out massive numbers of students to study abroad, to learn the best practices and bring them home. There will be an active push toward sustainable and green technologies. However, their primary driver will not be climate change, but the polluted megacities of China. The technologies in this field will be increasingly geared toward making human life bearable in large industrial megalopolises, which means that some of the vital environmental and climate-related problems will not be appropriately addressed.

Europe will be sandwiched between China and the US. Acting more like a field of competition for the two, Europe's grip on its destiny will be more tenuous now than before. Most of the consumer technology innovation will be done elsewhere, and the economy of the old continent will be sluggish, much due to the aging population. In general, large webscale companies, like Google, Amazon, Alibaba, and Baidu, will dominate the markets. Europe will not have been able to create its own tech giants, although it will try to regulate what is done in its market area.

Cyberthreats will be persistent and demand more funds, and the best talents will be sought after for superpower projects to develop cyber-capabilities. This will have led to increasing demands for privacy protection in Western countries. At the same time, the rights of citizens in many countries will be curtailed by omnipresent AI-powered surveillance. There will be a need for trust and security, especially in the Western markets. Cyberespionage and sabotage will take place continually, as will information operations. Individuals, companies, and government organizations will be constantly looking for ways to reduce and eliminate vulnerabilities in their systems, which will tie up resources and cause delays. Resilient technologies, referring to technologies that are both resilient to external damages (both natural and human-caused), while at the same time promote the resilience of nature, will have emerged as an important field. In the West, resilience will have partly replaced sustainability in political and research discourse.

The US Order: The US Remains the Dominant Power on the Globe (4D)

The impressive economic gains made by China until 2022 will have been followed by a slower pace, mainly because of its unfavorable age structure. The idea of Chinese dominance spreading over Asia, Europe and Africa will have slowly dropped from conversation as it will have become clear that it needs to focus on keeping its population both peaceful and reasonably prosperous: China will have started to turn inwards. Despite this, more than ever the world will be replete with disinformation, half-truths, and manipulated facts. The reason for this will be that those opposed to the US-led order will have understood that their only option is to attack in the only way they truly can challenge the dominant superpower: by conquering, or at least demoralizing, the minds of those on the other side. The aim will be to weaken the national unity of the

Western countries, break their mutual relations, and increase tensions between them. The large Western technology companies, followed by many others, will have implemented policies and measures to combat disinformation in conjunction with various universities. AI-powered tools, together with humans, will search and correct false information. At the same time, the governments of Europe and North America will have been increasingly engaged in strategic communication operations aimed at both their domestic and foreign audiences. Truth will have become elusive.

The data security questions which were a vital issue of the early 2020s will have been resolved to a degree as the largest US corporations, together with the government, will have implemented blockchain-based encryption for their communications and data storage. European companies will follow suit, but the EU will be hesitant to endorse the use of blockchain because of perceived possibilities for tax evasion and other harmful activities. Human-centeredness, or consumer-centeredness will have become the current driving force in technologies designed in the Western countries, meaning that more than ever, new consumer needs will be actively discovered and created through psychological, sociological, and cultural research. Some will have pointed out that these efforts, combined with the strategic communication operations, could be seen as active engineering of the minds of the people. However, the memory of the pandemic will have made people more trusting toward communication technology than ever. The slow introduction of autonomous vehicles and co-working between robots and humans will have kept manual laborer employed, but AI will have started to take over many white-collar jobs formerly held, for example, by lawyers and programmers.

Even though the coronavirus will have forced many locations to shift their entertainment offerings online, large concert crowds, full cinemas, and crowded restaurants returned quickly during early 2022. During the decade, the large-scale introduction of virtual/augmented reality (VR/AR) applications together with fast 5G networks and edge computing will have transformed some entertainment forms completely. Once the acute crisis of the early 2020s pandemic was over, attention will have turned again to keeping climate change at bay. Public subsidies and adventurous entrepreneurs will have resulted in some impressive gains in photovoltaics, wave power, and wood-based materials.

Discussion and Conclusions

Sixteen alternative future scenarios were developed under four embedded scenario logic themes: user/developer, firm, ecosystem, and geopolitics. The user/developer can be seen as a sub-set of firm-level, the firm is a sub-set of ecosystems, and geopolitics forms the widest contextual level for all scenarios. To summarize the discussion of 6G scenarios. First, the probability of the scenarios arising was evaluated against the identified forces influencing them. Next, the plausibility of the scenarios was assessed based on their coherence by examining the potential alternative futures for 6G business events that could occur within their assessment. The third assessment step was to identify which scenarios were the most preferred. The preferability assessments of the scenarios were based on the values and choices the teams made regarding alternative futures. Both the most probable and most plausible scenarios stem from evolutionary supply-driven trends toward a multi-local networked world based on strong trends with low anticipated uncertainty. The most probable and plausible business scenarios, OTT (2D) and MNO6.0 (2C), build on the balance between competition and protective market views. In the ecosystem themed scenarios, the multi-local scenario sharing both the dystopic and utopian themes can be seen as simultaneously the most probable and plausible. All the preferable scenarios, Gaia (3A), Edge (2A), and Customer6.0 (1A), represent revolutionary demand-driven transformations toward sustainability, empowerment, and open ecosystems building on the democratic geopolitical scenario. They are based on high impact forces with higher uncertainty compared to the most probable and plausible scenarios.

In the preferred 6G future, the automatic collection of different kinds of data from humans, our environment, and its analysis are used for highly sophisticated products and systems that make people's lives easier, more sustainable, and provide a better user experience through convenience. The edge resources will be operated by local communities expanding services to remote rural areas, or research organizations deploying their own edge resources to accelerate local innovation. 6G will enhance platform cooperatives, as well as the peer-to-peer economy and sharing economic models, and the progress of a human-driven fair data and developer economy as well as the fair distribution of wealth. To summarize, we identified drivers, barriers, and challenges regarding the choices for developing the preferred 6G business future. These drivers, barriers, and challenges could concern all stakeholders in future 6G business. Key transformative global drivers concern climate change, sustainable development goals, and decentralization toward a networked poly-nodal world. External barriers to preferable future scenarios involve the uncertainties related to the power of dominating platforms, AI and HMI rights, democracy, and the regulation of resources. Key internal challenges, such as building disruptive business models and leveraging sharing economy antecedents while coping with the empowered users' and developers' rights, were identified.

The scenarios presented open a multitude of alternative futures for 6G. We will summarize and discuss our findings by examining what kind of economic, societal, and environmental notions we may draw from the scenarios.

From the *economic* perspective, the key messages of the four sets of scenarios can be summarized as follows:

- User experience will be customized, and resource orchestration will become user and developer centric.
- Local-demand–supply-consumption models will become prominent in an already globalized world, with a marked emphasis on localized spatial circular economies.
- New societal models for future service provisioning will emerge building on community-driven networks and public–private partnerships.
- Platform-based ecosystems will not only offer search, social media, and ecommerce but will provide an infrastructure in which innovation and transaction platforms are built by the developers.
- A strong role for 6G in various vertical and industrial context will continue the 5G evolution.
- The decoupling of technology platforms will lower the market entry barrier, allowing multiple entities to contribute to the innovations envisaged with 6G.

- Fine-grained technological and service modularity and open source will allow highly specialized solutions and services from developers and smaller entities to be widely deployed.
- Decentralized platform cooperatives will become counterforces to winner-takes-all platform monopolies.

From the *societal* perspective, the scenarios indicate the following:

- Hybrid military, economic, technological, and cultural powers will overlap, exercising threats and hybrid influence.
- Tensions between competitive, protective, networked, and empowered worldviews will grow.
- The power configuration may be transforming from a multi-polarized world to a poly-nodal world in which power will be determined in economic, technological, and cultural networks and interaction.
- Empowering experiential citizens as knowledge producers, developers, and users will contribute to a process of human-centered democratizing innovation stemming from pluralism and diversity.
- Privacy regulation will be strongly linked to the rising trends of the platform data economy, sharing economy, intelligent assistants, connected living in smart cities, transhumanism, and digital twins' 'meta' reality.

From the *environmental* perspective:

- 6G is seen as a provider of services to help steer communities and countries toward reaching the UN SDGs.
- 6G will offer opportunities for monitoring and steering the circular economy and understanding the big picture of the sustainable data economy.
- In utilizing sharing and circular economy trends, co-creation partners will employ existing resources and processes to promote the sustainable interaction.
- Companies will shift their focus, developing products and technologies that innovate to zero, including zero-waste and zero-emission technologies bringing social innovation to the forefront.

- Immersive digital realities will facilitate novel ways of learning, understanding, and memorizing subjects in many sciences such as chemistry, physics, biology, medicine, and astronomy.

In this context, the mobile communications industry is one of the sectors that has been very visibly linked to most elements of global disruptions because of its criticality to economic competitiveness as well as its visible interlinkage to the emergent digital business models (Ahokangas et al., 2022; Kilkki et al., 2018; Yrjola et al., 2022). A well-known example in this regard is the clean network initiative in the US and Chinese telecommunication company Huawei which has been subject to bans and strict oversight in the US and different Western countries due to the concerns for privacy and security (e.g., Hoffmann et al., 2019). As the deployment of 5G and 6G research is taking shape globally, there is a race to set the standards and protect technological innovation, and geopolitical disruptions (and considerations) are a core aspect of this debate (Yrjola et al., 2020; Klement, 2021). At the same time, it is vital to stress that the influence of a variety of global disruptions on 6G development and planned implementation has not been studied specifically so far, thereby revealing a visible gap in the extant literature.

Bibliography

Agarwal, R., Bajada, C., Green, R., & Skellern, K. (Eds.). (2021). *The Routledge companion to global value chains: Reinterpreting and reimagining megatrends in the world economy.* Routledge.

Ahokangas, P., Matinmikko-Blue, M., Latva-aho, M., Seppänen, V., Arslan, A., & Koivumäki, T. (2022). Future mobile network operator business scenarios. In R. Baikady, S. Sajid, V. Nadesan, J. Przeperski, M.R. Islam, & J. Gao (Eds.), *The Palgrave handbook of global social change.* Palgrave Macmillan, Cham. https://doi.org/10.1007/978-3-030-87624-1_20-1.

Ali, I., Arslan, A., Chowdhury, M., Khan, Z., & Tarba, S. Y. (2022). Reimagining global food value chains through effective resilience to COVID-19 shocks and similar future events: A dynamic capability perspective. *Journal of Business Research, 141,* 1–12.

Bennett, N., & Lemoine, G. J. (2014). What a difference a word makes: Understanding threats to performance in a VUCA world. *Business Horizons, 57*(3), 311–317.

Buckley, P. (2019). *International business in a VUCA world: The changing role of states and firms.* Emerald Group Publishing.

Chin, W. (2019). Technology, war and the state: Past, present and future. *International Affairs, 95*(4), 765–783.

Cho, D. S., & Moon, H. C. (2022). *The competitiveness of nations: Navigating the US-China trade war and the Covid-19 global pandemic.* World Scientific.

Dieppe, A. (Ed.). (2021). *Global productivity: trends, drivers, and policies.* World Bank Publications.

Greene, J. R., Krouskos, S., Hood, J., Basnayake, H., & Casey, W. (2018). *The stress test every business needs: A capital agenda for confidently facing digital disruption, difficult investors, recessions and geopolitical threats.* John Wiley & Sons.

Hoffmann, S., Bradshaw, S., & Taylor, E. (2019). *Networks and geopolitics: How great power rivalries infected 5G.* Oxford Information Labs. https://oxil.uk/publications/geopolitics-of-5g/Geopolitics_5G_Final.pdf.

Ivanov, D., Dolgui, A., & Sokolov, B. (Eds.). (2019). *Handbook of ripple effects in the supply chain* (Vol. 276). Springer.

Ivanov, D. (2021). *Introduction to supply chain resilience: Management, modelling, technology.* Springer Nature.

Kilkki, K., Mäntylä, M., Karhu, K., Hämmäinen, H., & Ailisto, H. (2018). A disruption framework. *Technological Forecasting and Social Change, 129,* 275–284.

Kinnula, M., & Iivari, N. (2019). Empowered to make a change: Guidelines for empowering the young generation in and through digital technology design. *Proceedings of the FabLearn Europe 2019 Conference,* 1–8.

Klement, J. (2021). *Geo-Economics: The interplay between geopolitics, economics, and investments.* CFA Institute Research Foundation. Available online at https://www.cfainstitute.org/-/media/documents/book/rf-publication/2021/geo-economics-full.pdf.

Korn, T., & Stemmler, H. (2022). *Russia's war against Ukraine might persistently shift global supply chains. VoxEU.* https://cepr.org/voxeu/columns/russias-war-against-ukraine-might-persistently-shift-global-supply-chains.

Lukin, A. (2019). The US–China trade war and China's strategic future. *Survival, 61*(1), 23–50.

Millar, C. C., Groth, O., & Mahon, J. F. (2018). Management innovation in a VUCA world: Challenges and recommendations. *California Management Review, 61*(1), 5–14.

OECD (2022). *The supply of critical raw materials endangered by Russia's war on Ukraine.* https://www.oecd.org/ukraine-hub/policy-responses/the-supply-of-critical-raw-materials-endangered-by-russia-s-war-on-ukraine-e01ac7be/.

Ozbayrac, G. (2022). *Enterprise agility: A practical guide to agile business management.* CRC Press.

Petricevic, O., & Teece, D. J. (2019). The structural reshaping of globalization: Implications for strategic sectors, profiting from innovation, and the multinational enterprise. *Journal of International Business Studies, 50*(9), 1487–1512.

Schoemaker, P. J. (1995). Scenario planning: A tool for strategic thinking. *MIT Sloan Management Review.*

Yrjölä, S., Ahokangas, P., & Matinmikko-Blue M. (Eds.) (2020). *White Paper on Business of 6G.* 6G Research Visions, 3, University of Oulu, Oulu, Finland.

Yrjölä, S., Ahokangas, P., Arslan, A., Matinmikko-Blue, M., Golgeci, I., & Tarba, S. (2022). Artificial intelligence in the telecommunication sector: Exploratory analysis of 6G's potential for organizational agility. In *Entrepreneurial connectivity* (pp. 63–81). Springer.

6G Flagship. (2020). *6G Research Visions 2 [White paper].* University of Oulu.

4

Sustainability Transition and 6G Mobile Communications

Marja Matinmikko-Blue and Ahmad Arslan

Sustainable development is development that meets the needs of the present without compromising the ability of future generations to meet their own needs.

(Gro Harlem Brundtland)

M. Matinmikko-Blue (✉)
Infotech Oulu Focus Institute and Centre for Wireless Communications, University of Oulu, Oulu, Finland
e-mail: marja.matinmikko@oulu.fi

A. Arslan
Department of Marketing, Management, and International Business, Oulu Business School, University of Oulu, Oulu, Finland

Innolab, University of Vaasa, Vaasa, Finland

A. Arslan
e-mail: ahmad.arslan@oulu.fi

© The Author(s) 2024   **93**
P. Ahokangas and A. Aagaard (eds.), *The Changing World of Mobile Communications*,
https://doi.org/10.1007/978-3-031-33191-6_4

Introduction to Sustainability Transition

Sustainability has become a buzzword in academia as well as policy-making circles these days due to visible influences of climate change on daily life (e.g., Arslan et al., 2021; Barber, 2021; Heikkurinen & Ruuska, 2021) along with the recognition that social sustainability lies at the core of achieving the UN sustainable development goals (Baldwin & King, 2018; Ranjabari et al., 2021). Scholars have started to include sustainability principles in their research in various fields in response to growing sustainability concerns as well as funding agency require-ments. Consequently, the number of publications focusing on various aspects of sustainability in different industries, national contexts, and organizational settings has increased. In parallel, firms have also increas-ingly started to see sustainability not as an additional cost, but also as a business opportunity.

Elkington's (1997) environmental, social, and economic approaches toward sustainability is often referred to as the triple bottom line (TBL). To manage the interdependence, the demands stemming from the three perspectives should not be compromised but balanced. Economic sustainability aims to secure profitability and liquidity; social sustain-ability aims to contribute to human and social capital; and environ-mental sustainability favors the consumption of reproducible resources (Khan et al., 2021).

In recent discussions, the term resilience has started to emerge as connected with sustainability, albeit without clarity regarding the difference between the two. The extant literature views resilience and sustainability either as: (1) independent and separate; (2) overlapping or complementary; or (3) a component of the other (Marchese et al., 2018). Resilience refers to the ability of an entity or system to return to a normal condition after disruption—being a measure of a system's ability to absorb continuous and unpredictable change and continue to function (Hosseini et al., 2016; Pregenzer, 2011). Thus, the term value can be found at the hearth of both sustainability and resilience, i.e., to identify, create, convey, deliver, and capture, but also protect and sustain long-term value, whether economic, environmental, or social (Liu et al., 2021).

Sustainability, Innovation, and Disruption

In the context of sustainability, there is an increasing recognition among management scholars that understanding the transition toward environmental and social sustainability is vital despite attractive slogans. If the actual sustainability transition process is not understood and managed well, achieving sustainable development-related goals will be harder (Bai et al., 2009; Geels, 2011; Heikkurinen & Ruuska, 2021; Williams & Robinson, 2020). This sustainability transition is highly linked to economic sustainability, which has limited the actions on environmental and social sustainability. In this context of the sustainability transition, the role of disruptive innovations has emerged as critical in recent years because these disruptive innovations demand socio-technical change at multiple levels (Heikkurinen & Ruuska, 2021; Park et al., 2021); thereby bringing the transition element to the forefront of the debate (e.g., Bai et al., 2009; Brauch et al., 2016; Geels, 2011, 2019; Kivimaa et al., 2021). This sustainability transition approach is different from many traditional sustainability focused studies, which either focus on a micro-context (firm level sustainability initiatives) or macro-level (change toward sustainability in industries and countries), where the process of this transition does not usually get the due attention.

The sustainability transition has primarily been studied in the context of innovation in the energy sector due to its' visible linkages with environmental degradation (e.g., Bogdanov et al., 2021; Brauch et al., 2016; Kivimaa et al., 2021). However, calls have been made by scholars to apply a wider approach to studying sustainability transition in relation to innovations (disruptive innovations) in different industries, national contexts, and organizational settings (e.g., Rohe & Chlebna, 2022; van der Loos et al., 2020, 2022). At the same time, the other critical element of the sustainability transition associated with social sustainability is even less studied, and most research in the larger field of management has focused on organizational responsibilities and policy initiatives leading to social sustainability, so far (e.g., Hutchins & Sutherland, 2008; Amrutha & Geeta, 2020; Ranjbari et al., 2021). A review of the prior literature further reveals that the potential of disruptive innovations in the transition toward social sustainability is rarely studied; a visible gap that our chapter aims to fill is the potential of 6G mobile telecommunications technology.

6G systems have a high potential to contribute to both environmental and social sustainability while ensuring economic sustainability, and this has been established by several studies published in recent years (e.g., Matinmikko-Blue et al., 2020; Matinmikko-Blue et al., 2021; Ojutkangas et al., 2022). However, as 6G is still a future technology in the vision and framework development phase, we still lack knowledge of how it can potentially contribute to the sustainability transition on environmental, economic, and social levels. Prior work has linked 6G with the UN SDGs (Matinmikko-Blue et al., 2020; Ojutkangas et al., 2022) and the triple bottom line of sustainability (Matinmikko-Blue et al., 2021, 2022) and identified several research topics for further study by the research community including environmental, economic, and social perspectives.

Sustainability considerations of existing mobile communication systems have primarily focused on environmental sustainability aiming at minimizing energy consumption and maximizing resource efficiency including energy efficiency (Zhang et al., 2016). The role of mobile communication is seen as important in the sustainability transition of society at large (Wu et al., 2018), but this development should not occur at the expense of increasing the ICT sector's own sustainability burden. Most recently, sustainability has become an important design criterion for 6G, (ITU-R, 2022; Matinmikko-Blue et al., 2020), while opening the door for defining a new set of requirements on mobile communications stemming from the sustainability transition.

Aims of the Chapter

The current chapter aims to address the sustainability transition and 6G interlinkage conceptually along with substantiating the discussion with some practical examples, using the most prominent approach used in transition studies, i.e., a multi-level perspective (MLP) (Geels, 2002, 2011, 2019, 2020; Rip & Kemp, 1998), which combines ideas from innovation studies, sociology, evolutionary economics, and institutional theory. The core argument of MLP is that transition is a result of a dynamic process at three different levels including: (1) niches, which are the core premises where the radical innovations are developed; (2) socio-technical regimes, representing institutional drivers toward the change; and (3) the exogenous socio-technical landscape of the larger society. By establishing the link between 6G and the sustainability transition including environmental, economic, and social perspectives using the MLP lens, our chapter offers two critical contributions to the extant 6G, sustainability transition, and innovation management literature streams. Firstly, this chapter highlights the potential of 6G in both environmental and social sustainability conceptually while ensuring economic sustainability as well as practically by referring to examples. Secondly, it is one of the rare studies that focuses on the larger picture in the 6G and sustainability debate by highlighting specific UN SDGs which can be achieved in the sustainability transition and the role of endogenous and exogenous factors using the MLP lens. Hence, the potential practical and policy implications of this chapter are expected to be profound.

The rest of this chapter is organized in the following manner. The next section analyzes the established connection between sustainability and 6G. After that, 6G development in relation to the sustainability transition is analyzed via the MLP approach. Here, the niche aspects are presented, after which the socio-technical regimes and institutional factors linked to the sustainability transition in this context are discussed along with the wider debate on the UN SDGs. The sub-section after that discussion aims to bring the larger society into debate by focusing on exogenous the socio-technical landscape of 6G development and sustainability transition. The last section presents theoretical and policy implications, along with a discussion on the study limitations and future research directions, restructuring possibilities linked to 6G are discussed in relation to the increased complexity of the external environment.

Connecting Sustainability to 6G

6G mobile communication systems are expected to be deployed around the year 2030, which is also the target year for the achievement of the UN SDGs (Matinmikko-Blue et al., 2020). Idealistically, the targets from the UN SDG framework should be reached prior to the emergence of 6G, allowing 6G to enter a world where major sustainability challenges are already solved. This, however, will not be the case and the entire R&D of the next generation of mobile communication systems is driven by sustainability and sustainable development (Latva-aho & Leppänen, 2019; Matinmikko-Blue et al., 2020). No prior generation of mobile communications has taken sustainability as seriously as a core value as 6G has. The development of the 4G system adopted the principle of green communications (Zhang et al., 2016), which meant optimization of resource usage and especially energy efficiency. 5G adopted energy efficiency as one of its key performance indicators (ITU-R, 2017), but no target values were defined. In 6G development, sustainability principles are talked about but concrete actions and design criteria for sustainability in 6G R&D are yet un(der)defined.

The triple bottom line of sustainability (Elkington, 1997) has highlighted the interdependencies between social, environmental, and economic sustainability principles. This also applies to 6G, where the ongoing 6G development aims at solving major social and/or environmental sustainability challenges, while being economically sustainable. The attempts to enable the sustainability transition in other sectors of society through the so-called enablement effect has received considerable attention (Hilty & Aebischer, 2015). The quantification of the enablement effect in terms of the reduction of greenhouse gas emissions (ITU-T, 2022) or energy consumption, for example, has turned out to be a challenge both from methodological and practical perspectives. At the same time, understanding the mobile communication sector's own sustainability impact including lifecycle greenhouse gas emissions (ITU-T, 2018) is a challenging topic as well, and reported data remains low. Both the enablement effect and the mobile communications' own sustainability burden remain equally important in the sustainability transition but require distinct methods and measures.

The original connection between 6G and the UN SDGs was developed in the 6G Flagship's white paper work (Matinmikko-Blue et al., 2020), which stressed a three-fold role for 6G.

- Firstly, 6G will be a provider of services that help steer communities and nations toward the UN SDGs. As an example, global mobile connectivity that will connect the unconnected is a foreseen scenario for 6G requiring cost-efficient and deployable solutions.
- Secondly, 6G will become a powerful measurement tool for data collection at a very local level of granularity to help organizations and nations to report on sustainability-related indicators, which today is a problem. Examples include sensing solutions for collecting environmental data.
- Thirdly, 6G will be a reinforcer of a new technological ecosystem which will be developed according to the high-level requirements set in the UN SDGs.

As an example, translating the requirements stemming from the UN SDG framework in areas such as promoting high-quality education (SDG4), promoting equality (SDG5, SDG10) and digital inclusions (SDG9), to 6G technology development requires introducing new design goals that are both technical and regulatory in nature. The existing indicators in the UN SDG framework address SDG4 (the proportion of schools with access to the Internet for pedagogical purposes; the proportion of schools with access to computers for pedagogical purposes; the proportion of youth/adults with ICT skills, by type of skills), SDG5 (the proportion of individuals who own a mobile telephone, by sex; the percentage of the population covered by a mobile network, broken down by technology) and SDG17 (fixed Internet broadband subscriptions, broken down by speed; the proportion of individuals using the Internet) at a high level. For a sustainability transition where 6G is designed to contribute to the achievement of sustainability targets, contributions to other indicators are also expected as outlined in (Matinmikko-Blue et al., 2020).

A Multi-Level Perspective on the Sustainability Transition in 6G

Using the MLP lens, we analyze sustainability transition in the context of 6G. Our analysis stems from the pioneering work done in the 6G Flagship white paperwork (Latva-aho & Leppänen, 2019; Matinmikko-Blue et al., 2020), followed by several stakeholder interactions at workshops organized in 2021–2022.

Niche Aspects

Niches representing the core premises, where the radical innovations are developed, constitute of two factors in the context of 6G. Firstly, innovations in the context of the sustainability transition in 6G target the enablement effect of using innovative 6G service and solutions to help other sectors of society to act upon their sustainability impact toward environmental and/or social sustainability while maintaining economic sustainability. These innovations reduce GHG emissions and resource consumption including energy consumption through various means. Examples include optimizing processes in an industrial setting, reduction of fuel consumption in a multi-stakeholder port ecosystem with situational awareness, and productivity improvements in agriculture via sensors. Secondly, innovations within the 6G context will take place in 6G solutions and services to improve 6G's environmental and/or social sustainability while ensuring profitability in economic sustainability. Examples of these innovations include architectural designs to optimize the location of computing and communication resources, algorithms to minimize energy consumption in different layers of the network, and cost-efficient network deployment models in challenge areas.

The dual role of ICTs and particularly mobile communications in helping other sectors to achieve sustainability-related targets as well as the mobile communication sector's own sustainability impact are two sides of the complex sustainability transition challenge. Agreed metrics and methods are needed for both sides as presented in (ITU-T, 2022) and (ITU-T, 2018), respectively.

Socio-Technical Regimes and Institutional Factors

Socio-technical regimes and institutional factors in the context of 6G particularly include regulations and standardization, which are discussed in more detail other chapters of this book. The UN SDG framework presented in Agenda 2030 (UN, 2015) is an international treaty that nations have agreed to. At regional level in Europe, the European Commissions' Green Deal positions Europe's targets for member states. Consequently, national level approaches in the EU member countries follow both international and European level rules.

Traditionally, environmental sustainability is the matter of the ministry of environment, while telecommunication communications and services typically belong to the ministry of communications. The introduction of the topics of sustainability of the ICT sector and the enablement effect of the ICT sector in other sectors have resulted in a new setting, where traditional governance boundaries are broken. Consequently, the expertise required for environmental sustainability is not found in traditional telecommunications.

At the European level, the European telecommunications regulators at BEREC have stated that sustainability is a new topic for them (BEREC, 2022). Prior studies on the environmental sustainability of the ICT sector have resulted in conflicting results. One challenge is the lack of data available for unbiased research by the research community. In their current work on sustainability of the ICT sector, much focus is being put on mobile network operators (MNOs) and infrastructure vendors' views.

Governments set sustainability-related requirements in the different sectors of society. Until now, there have not been specific sustainability requirements for prior generations of mobile communication networks. On the one hand, coverage obligations for mobile communication networks introduced early on in Finland and in some other countries can be seen as an early form of a social sustainability requirement. These obligations mandate the MNOs to deploy networks in geographical areas that cover a certain percentage of the population such as 99%. Concrete sustainability requirements for 6G defined by governments are not yet known. International-level discussions have started at the UN-based agency for ICT, the International Telecommunication Union (ITU), and specific requirements for the radio interface will be defined in the coming years.

The Exogenous Socio-Technical Landscape

The wider society impacting or being impacted by 6G in the sustainability transition particularly involves end users. They have the power to decide on the use of ICT services and solutions. Today, end users do not have information available about the sustainability impact of their ICT choices. In the future, end users will be able to make informed decisions based on sustainability-related information. For example, the environmental footprint of the service usage including energy consumption, greenhouse gas production, and other depletion of natural resources will impact consumer behavior once this information is made available. Today, this is not possible due to the lack of information for consumers and other end users.

More widely, the inclusion of a human perspective in 6G development in the sustainability transition requires the introduction of proper stakeholders in the R&D process at the right stages. Traditionally, mobile communication systems have been defined by the companies involved in the development of the technology and the governments, emphasizing high-level requirements such international roaming, and moving of equipment as well as detailed technical requirements on aspects such as capacity and delays. The research and development of the systems is carried out by the technology vendors in close collaboration with the research domain aiming to respond to the requirements of the customers, who in the mobile communication business are typically MNOs and their end users. Within these requirements, sustainability-related requirements are only emerging, and their format is not clear yet. Additionally, regulators play a key role in setting these requirements. In the deployment and operational phase, the role of the MNOs is critical as they run the infrastructure and act as the interface for the end users. The reduction of energy consumption is already a top requirement by the MNOs to reduce their operational costs and will continue to be so in 6G.

The above discussion also connects to the discussion on system resilience in 6G that covers resilience to cyber-attacks and the utilization of privacy-preserving and trust-creating technologies to achieve trustworthy 6G. Trustworthy 6G comprises topics such as security, privacy, safety, availability, inclusivity, transparency, fairness, accountability, resilience, and compliance with ethical frameworks. In addition, as with artificial intelligence, human agency and oversight based on values should be considered in 6G. These topics correspond with the emerging social sustainability and resilience themes even toward sovereignty and democracy.

Theoretical and Policy Implications

Our study on the sustainability transition in 6G as analyzed via the multi-level perspective offers both theoretical and policy implications. Firstly, a major theoretical implication relates to the specificity of the 6G context in relation to the sustainability transition and the need for theorization. The role of telecommunications services has evolved from offering organization communication tools to multi-level influences across the socioeconomic and technical landscape. The discussion presented in this chapter has shown that 6G can potentially ensure more environmental and social sustainability which can ultimately lead to economic sustainability. This interlinkage has all of the elements of the TBL (triple bottom line) view of sustainability, but at the same time it depicts the criticality of social and environmental sustainability in ensuring economic sustainability. Hence, this aspect should be further explored and theorized in the specific context of telecom sector and 6G as it will enrich the sustainability debate in a novel way. Secondly, the current chapter enriches the understanding of the UN SDGs by linking them with all three levels of MLP concerning the sustainability transition. Hence, we set the bases for future studies to undertake a more in-depth approach to analysis and offer a fine-grained view of these interlinkages in different industrial settings as well as geographical contexts.

On the policy side, regulations are expected to emerge to regulate 6G including the sustainability and resilience perspectives. Traditional requirements for mobile communication systems need to be complemented with a thorough sustainability transition perspective and those discussions have only just started. Today, the future upcoming regulations are unknown, and their derivation requires new skill combinations bringing together telecommunications, environmental, and social and economic sustainability knowhow.

In particular, regulators and policymakers at different levels from the community, national, regional to the international level, will face a new situation with requirements that do not exist today. Yet, national-level approaches should not be too different from each other to allow economic sustainability of the entire mobile communication sector where mobility across borders has been the initial driver. The global success of mobile communications has been based on global harmonization and roaming of devices leading to large markets for the same devices, which has brought prices down and allowed the same equipment to work in a number of countries. If the sustainability requirements are different from country to country, fragmentation of equipment may lead to markets not being served or prices becoming high.

Limitations and Future Research Directions

Our chapter has several limitations similar to any other academic work. Firstly, it is a conceptual piece where empirical analysis has not been undertaken. However, as 6G is a future technology, the possibilities for a specific analysis of its link to the sustainability transition are rather limited. Hence, our chapter builds bases for future studies to be undertaken both as quantitative and qualitative studies analyzing different aspects of the sustainability transition in relation to 6G telecommunications in different industrial and national contexts. Moreover, our chapter discusses all three elements of MLP concerning the sustainability transition without going into too much depth on any of the elements. We recommend future scholars take a more in-depth approach and analyze the specificities of the various niches, socio-technical regimes, and the exogenous socio-technical landscape of the sustainability transition in different contexts in relation to 6G telecommunications. Additionally, keeping in view the importance of corporate social and environmental innovation (e.g., Golgeci et al., 2022) for the sustainability transition, we recommend future researchers to link these to 6G telecoms as well; thereby enriching the larger debate on the sustainability transition linked to this particular technology. Finally, keeping in view the continuous development of 6G telecoms currently taking place, longitudinal academic studies documenting different phases in relation to the sustainability transition are expected to enrich our understanding both theoretically and practically.

Bibliography

Amrutha, V. N., & Geetha, S. N. (2020). A systematic review on green human resource management: Implications for social sustainability. *Journal of Cleaner Production, 247*, 119131.

Arslan, A., Haapanen, L., Hurmelinna-Laukkanen, P., Tarba, S. Y., & Alon, I. (2021). Climate change, consumer lifestyles and legitimation strategies of sustainability-oriented firms. *European Management Journal, 39*(6), 720–730.

Bai, X., Wieczorek, A. J., Kaneko, S., Lisson, S., & Contreras, A. (2009). Enabling sustainability transitions in Asia: The importance of vertical and horizontal linkages. *Technological Forecasting and Social Change, 76*(2), 255–266.

Baldwin, C., & King, R. (2018). *Social sustainability, climate resilience and community-based urban development: What about the people?* Routledge.

Barber, A. (2021). *Consumed: The need for collective change; colonialism, climate change & consumerism.* Hachette.

BEREC. (2022). BEREC *Report on sustainability: Assessing BEREC's contribution to limiting the impact of the digital sector on the environment.* https://www.berec.europa.eu/system/files/2022-07/10282-berec-report-on-sustainability-assessing_0_3.pdf.

Bogdanov, D., Gulagi, A., Fasihi, M., & Breyer, C. (2021). Full energy sector transition towards 100% renewable energy supply: Integrating power, heat, transport and industry sectors including desalination. *Applied Energy, 283,* 116273.

Brauch, H. G., Spring, Ú. O., Grin, J., & Scheffran, J. (Eds.). (2016). *Handbook on sustainability transition and sustainable peace.* Springer.

Elkington, J. (1997). *Cannibals with forks: The triple bottom line of 21st-century business.* Capstone Publishing Ltd.

Geels, F. W. (2002). Technological transitions as evolutionary reconfiguration processes: A multi-level perspective and a case-study. *Research Policy, 31*(8–9), 1257–1274.

Geels, F. W. (2011). The multi-level perspective on sustainability transitions: Responses to seven criticisms. *Environmental Innovation and Societal Transitions, 1*(1), 24–40.

Geels, F. W. (2019). Socio-technical transitions to sustainability: A review of criticisms and elaborations of the multi-level perspective. *Current Opinion in Environmental Sustainability, 39,* 187–201.

Geels, F. W. (2020). Micro-foundations of the multi-level perspective on socio-technical transitions: Developing a multi-dimensional model of agency through crossovers between social constructivism, evolutionary economics and neo-institutional theory. *Technological Forecasting and Social Change, 152,* 119894.

Golgeci, I., Ali, I., Bozkurt, S., Gligor, D. M., & Arslan, A. (2022). The impact of corporate support programs on environmental and social innovation: Empirical insights from the food and beverage industry. *International Journal of Operations & Production Management, 42*(7), 898–929.

Heikkurinen, P., & Ruuska, T. (2021). *Sustainability beyond technology: Philosophy, critique, and implications for human organization.* Oxford University Press.

Hilty, L. M., & Aebischer, B. (2015). ICT for sustainability: An emerging research field. In Hilty, L., Aebischer, B. (Eds.), ICT innovations for sustainability. *Advances in intelligent systems and computing,* Vol. 310. Springer, Cham. https://doi.org/10.1007/978-3-319-09228-7_1.

Hosseini, S., Barker, K., & Ramirez-Marquez, J. E. (2016). A review of definitions and measures of system resilience. *Reliability Engineering & System Safety, 145,* 47–61.

Hutchins, M. J., & Sutherland, J. W. (2008). An exploration of measures of social sustainability and their application to supply chain decisions. *Journal of Cleaner Production, 16*(15), 1688–1698.

International Telecommunication Union Radiocommunication sector (ITU-R). (2017). *Minimum requirements related to technical performance for*

IMT-2020 radio interface(s). Report ITU-R M.2410. https://www.itu.int/pub/R-REP-M.2410-2017.

International Telecommunication Union Radiocommunication sector (ITU-R). (2022). *Future technology trends of terrestrial IMT systems towards 2030 and beyond.* Report ITU-R M.2516. https://www.itu.int/pub/R-REP-M.2516-2022.

International Telecommunication Union Standardization sector (ITU-T). (2018). *Methodologies for the assessment of the environmental impact of the information and communication technology sector.* Recommendation ITU-T L.1450. Available online at https://www.itu.int/rec/T-REC-L.1450-201809-I/en.

International Telecommunication Union Standardization sector (ITU-T). (2022). *Enabling the Net Zero transition: Assessing how the use of ICT solutions impacts GHG emissions of other sectors.* Recommendation ITU-T L.1480. https://www.itu.int/ITU-T/workprog/wp_item.aspx?isn=17698.

Khan, I. S., Ahmad, M. O., & Majava, J. (2021). Industry 4.0 and sustainable development: A systematic mapping of triple bottom line, circular economy and sustainable business models perspectives. *Journal of Cleaner Production, 297,* 126655.

Kivimaa, P., Laakso, S., Lonkila, A., & Kaljonen, M. (2021). Moving beyond disruptive innovation: A review of disruption in sustainability transitions. *Environmental Innovation and Societal Transitions, 38,* 110–126.

Latva-aho, M., & Leppänen, K. (Eds.). (2019). *Key drivers and research challenges for 6G ubiquitous wireless intelligence.* [White paper]. http://urn.fi/urn:isbn:9789526223544.

Liu, J., Tong, T. W., & Sinfield, J. V. (2021). Toward a resilient complex adaptive system view of business models. *Long Range Planning, 54*(3), 102030.

Marchese, D., Reynolds, E., Bates, M. E., Morgan, H., Clark, S. S., & Linkov, I. (2018). Resilience and sustainability: Similarities and differences in environmental management applications. *Science of the Total Environment, 613,* 1275–1283.

Matinmikko-Blue et al. (Eds.). (2020). *6G Drivers and the UN SDGs* [White paper]. http://urn.fi/urn:isbn:9789526226699.

Matinmikko-Blue, M., Yrjölä, S., Ahokangas. P. (2022). Sustainable technology design for future 6G mobile communications. In Leal Filho, W., Azul, A. M., Doni, F., Salvia, A. L. (Eds.), *Handbook of sustainability science in the future.* Springer, Cham. https://doi.org/10.1007/978-3-030-68074-9_1 49-1.

Matinmikko-Blue, M., Yrjölä, S., Ahokangas, P., Ojutkangas, K., & Rossi, E. (2021). 6G and the UN SDGs: Where is the connection? *Wireless Personal Communications, 121*(2), 1339–1360.

Ojutkangas, K., Rossi, E., & Matinmikko-Blue, M. (2022). A deep dive into the birth process of linking 6G and the UN SDGs. *Telecommunications Policy, 46*(1), 102283.

Park, S. H., Gonzalez-Perez, M. A., & Floriani, D. E. (Eds.). (2021). *The Palgrave handbook of corporate sustainability in the digital era.* Palgrave Macmillan.

Pregenzer, A. L. (2011). *Systems resilience: A new analytical framework for nuclear nonproliferation* (No. SAND2011–9463). Sandia National Laboratories.

Ranjbari, M., Esfandabadi, Z. S., Zanetti, M. C., Scagnelli, S. D., Siebers, P. O., Aghbashlo, M., … & Tabatabaei, M. (2021). Three pillars of sustainability in the wake of COVID-19: A systematic review and future research agenda for sustainable development. *Journal of Cleaner Production, 297*, 126660.

Rip, A., & Kemp, R. (1998). Technological change. *Human Choice and Climate Change, 2*(2), 327–399.

Rohe, S., & Chlebna, C. (2022). The evolving role of networking organizations in advanced sustainability transitions. *Technological Forecasting and Social Change, 183*, 121916.

United Nations. (2015). *Agenda 2030. Resolution adopted by the general assembly on transforming our world: The 2030 Agenda for Sustainable Development.* (A/RES/70/1). United Nations.

van der Loos, H. A., Negro, S. O., & Hekkert, M. P. (2020). International markets and technological innovation systems: The case of offshore wind. *Environmental Innovation and Societal Transitions, 34*, 121–138.

van der Loos, A., Langeveld, R., Hekkert, M., Negro, S., & Truffer, B. (2022). Developing local industries and global value chains: The case of offshore wind. *Technological Forecasting and Social Change, 174*, 121248.

Williams, S., & Robinson, J. (2020). Measuring sustainability: An evaluation framework for sustainability transition experiments. *Environmental Science & Policy, 103*, 58–66.

Wu, J., Guo, S., Huang, H., Liu, W., & Xiang, Y. (2018). Information and communications technologies for sustainable development goals: State-of-the-art, needs and perspectives. *IEEE Communications Surveys & Tutorials, 20*(3), 2389–2406.

Zhang, S., Wu, Q., Xu, S., & Li, G. Y. (2016). Fundamental green tradeoffs: Progresses, challenges, and impacts on 5G networks. *IEEE Communications Surveys & Tutorials, 19*(1), 33–56.

Part II

Value Creation and Capture in Future Mobile Communications

5

Value Creation and Services in Mobile Communications

Annabeth Aagaard, Petri Ahokangas, Marika Iivari, Irina Atkova, Seppo Yrjölä, and Marja Matinmikko-Blue

We cannot solve our problems with the same thinking we used when we created them.

(Albert Einstein)

Toward Sustainable Value Creation

Developments in Information and Communication Technologies (ICT) like mobile communications have substantially transformed service systems and extended the boundaries of service interactions, while providing new business opportunities for value creation (Edvardsson et al., 2018). In parallel, the traditional view that value is first created, then delivered, and finally captured by a focal firm has changed to a view in which value is simultaneously co-created and co-captured by and

A. Aagaard (✉)
Department of Management, Aarhus University, Aarhus, Denmark
e-mail: aaa@mgmt.au.dk

© The Author(s) 2024   **113**
P. Ahokangas and A. Aagaard (eds.), *The Changing World of Mobile Communications*,
https://doi.org/10.1007/978-3-031-33191-6_5

for both service providers and users in an ecosystem of actors. In practice, value co-creation and co-production are becoming more widespread due to ICT's capability to drastically reduce actors' coordination costs in co-producing services (Kallinikos, 2011). In short, this makes service processes interactive, collaborative, and able to involve multiple actors or service system entities (Vargo et al., 2010).

Successful value co-creation requires that actors can interact through exchange of resources, while integrating these in the context of their own ecosystemic business reality (Prahalad & Ramaswamy, 2004). Ecosystem actors are enabled to exchange resources to accumulate value, co-creating value increasingly through virtual, i.e., digital, rather than physical interfaces (Davis et al., 2011).

Many scholars extend the value construct from a one-dimensional shareholder logic of profit maximization to more stakeholders and levels of attention (Aagaard & Ritzén, 2020). In sustainable businesses, value

---

P. Ahokangas · M. Iivari
Martti Ahtisaari Institute, Oulu Business School, University of Oulu, Oulu, Finland
e-mail: petri.ahokangas@oulu.fi

M. Iivari
e-mail: marika.iivari@oulu.fi

I. Atkova
Martti Ahtisaari Institute, University of Oulu, Oulu, Finland
e-mail: Irina.atkova@oulu.fi

S. Yrjölä
Centre for Wireless Communications, University of Oulu, Oulu, Finland

Nokia, Oulu, Finland

S. Yrjölä
e-mail: seppo.yrjola@oulu.fi; seppo.yrjola@nokia.com

M. Matinmikko-Blue
Infotech Oulu Focus Institute and Centre for Wireless Communications, University of Oulu, Oulu, Finland
e-mail: marja.matinmikko@oulu.fi

propositions go far beyond economic considerations, and include environmental and social considerations, referred to as the triple bottom line logic (Bocken et al., 2015; Pedersen et al., 2018). Thus, all actors and entities should benefit from the transformation of value-in-use into value-in-exchange, meaning that value may be captured on more levels (Lepak et al., 2007). Sustainable value may be defined as *"a promise on the economic, environmental and social benefits that a firm's offering delivers to customers and society at large, considering both short-term profits and long-term sustainability"* (Patala et al., 2016, p. 1). This implies that value is embedded in a specific social context and consequently is defined differently by different players (Barrett et al., 2016).

It follows that an organizational value proposition varies depending on the contextual specifics or a 'regime of value' (Appadurai, 1986). A regime of value *"is a socially coherent and situated way of establishing what is valuable"* (Barrett et al., 2016, p. 709). Resonating with the idea of sustainability, and triple bottom line, a regime of value includes social, political, and economic regimes of value (Grover & Kohli, 2012). It is important to note that the existence of multiple value propositions can naturally lead to various tensions that, however, can be creatively leveraged to achieve synergies (Stark, 2009).

In business model research—that constitutes the mainstream of the value creation discussion today—value is closely intertwined with the concepts of opportunity and competitive advantage (Zott et al., 2011). Given the growing sustainability pressures, these concepts are increasingly discussed in the context of sustainability (Greissdoerfer et al., 2018). A possibility of the concurrent presence of several value propositions by different stakeholders in an ecosystem may entail variation in opportunity and competitive advantage conceptualizations, as well as in the processes of value creation (Corvellec & Hultman, 2014). Traditionally, business model discussions identify four sources of value creation: novelty, lock-in, complementarities, and efficiency (Amit & Zott, 2012). In this, novelty may serve as a starting point, with complementary elements to existing products and services leading to efficiency, or having it as a key feature of the novel offering, finally resulting in customer lock-in.

For 5G services, the basis for value creation comes from the three usage scenarios defined by ITU-R (2015) are as follows:

- Enhanced mobile broadband (eMBB) addresses human-centric use cases for access to multi-media content, services, and data as an extension to the mobile broadband services offered by earlier generations of mobile communications. The aim is to achieve improved performance and an increasingly seamless user experience both in hotspots and wide area coverage.
- Ultra-reliable and low-latency communications (URLLC) set stringent requirements for machine-centric use cases in terms of throughput, latency, and availability in various application areas including industry, healthcare, energy, or transportation.
- Massive machine-type communications (mMTC) are characterized by a high number of connected devices with relatively low volume of non-delay-sensitive data. These kinds of communications are typical for different Internet of Things (IoT) applications.

With these original usage scenarios, ITU-R has adopted a service-centric approach to value creation with mobile communications services. In practice, however, the value creation in mobile services has taken different forms. Drawing on the insights of extant research, this chapter approaches and conceptualizes opportunities for value co-creation and co-capture based on 5G and 6G usage scenarios as configurations on service, platform, and ecosystem levels of analysis (Grover & Kohli, 2012; Orlikowski & Scott, 2008). The purpose of the chapter is to provide a preliminary understanding of how the 5G/6G-enabled services may unfold in future, and what value they may bring to the evolving mobile communications ecosystem.

The chapter starts by providing a conceptual understanding of value creation in via services within platforms and ecosystems. What follows is a presentation of the current understanding of services within 5G focusing on different deployment modes. Since 6G is still in the research phase, the chapter discusses the capabilities and use case candidates identified for 6G as possible sources of value creation. The chapter concludes with a discussion on trustworthy 6G and outlook to applications and services in future 6G.

Digitalization, Platformization, and Ecosystems

Digitalization has enabled companies to create novel value and offer service configurations by incorporating and combining emerging technologies either as a main value driver or as a complementary agent (Parida et al., 2019). Reflecting on mobile communications and ICT in general, the business value creation has changed from a connectivity approach (ICT as a communication channel) through an immersion approach (ICT as an operating environment) to a fusion approach (ICT as fabric), where modular platforms can be adapted and interconnected in different ways (El Sawy & Pereira, 2013). Digital technologies have enabled opportunities for value co-creation through services by connecting actors who would otherwise not have been connected. Consequently, companies have increasingly adopted digital platforms to mitigate service innovation challenges (Rai et al., 2019). Through digital platforms, the flow of information is mediated, which enables the interconnection of products and services, as well as data flows between different actors (e.g., service providers, customers, and end users) on multiple sides of the platform (Ruutu et al., 2017). As an example, Apple and Google have established digital platforms to develop new mobile data services by acquiring commercial ideas from firm customers and consumers (Ye & Kankanhalli, 2018). However, in these technology-mediated ecosystems, value creation by itself is insufficient, as one needs to consider the extent to which agents can capture value in a fair way with respect to their contribution (Corsaro, 2020).

In the digital service context, defining and identifying individual contributions are not always straightforward, as some service processes might be hidden, and benefits and costs might be intangible (Hofacker & Corsaro, 2020). Furthermore, the value creation of digital services depends crucially on the network effects (Zhu & Iansiti, 2019) created by the platform. For example, the value of a mobile phone operating system platform for end users depends on the number of application developers as well as the applications developed by them (Garcia-Swartz & Garcia-Vicente, 2015). In sum, new technologies and the concomitant emergence of various digital services have enabled new means of value co-creation.

From Platformization…

The integration of various forms of value co-creation present requires ecosystem actors to interact through the exchange of resources in the context of their own reality (Prahalad & Ramaswamy, 2004). Platformization and the modularization facilitated by the convergence of ICT technologies enable novel configurations of needs and resources. It has been argued that platform-based digital markets can alter the way companies generate and deliver value to end customers (Cusumano et al., 2020). The purpose of platforms thus is to *"facilitate the multi-party exchange of products, which can be goods, services, or even social currency"* (Sorri et al., 2019, p. 2), in the creation of novel value while ensuring value capture. McIntyre and Srinivasan (2017, p. 143) state that *"platforms can be conceptualized as interfaces—often embodied in products, services, or technologies—that can serve to mediate transactions between two or more sides."* From a more technology-centric perspective, digital platforms can therefore be explained as software-based external platforms. These platforms consist of an extensible codebase of a software-based system, which provides core functionalities shared by the modules that are interoperating with it and the interfaces through which they are interoperating (Sorri et al., 2019).

Thus, in ecosystems characterized by technological interdependence between companies, the value created by a particular firm's technological choices is highly dependent on the choices made by other ecosystem actors that possess complementary technologies (Adner & Kapoor, 2010). Therefore, to unlock network effects and create 'lock-in,' technology ecosystem actors need to coordinate these choices not only with upstream and downstream value chain participants, but also with competitors and complementors (Vasudeva et al., 2020). Consequently, to fully leverage technological ecosystems aimed at value co-creation, adaptations are required by the complementors as well as the focal firm, resulting in a co-specialization of the underlying economic, technological, and cognitive architecture (Aarikka-Stenroos & Ritala, 2017; Autio & Thomas, 2019; Shipilov & Gawer, 2020). Furthermore, technology ecosystems constantly co-evolve and self-organize, leading to changes over time in the interdependencies and relationships between actors, co-creating value (Phillips & Ritala, 2019) and therefore require agile governance models.

... To Ecosystems

From a firm-centric perspective, value creation may be defined as an actor's attempt to increase value, whereas value capture is explained as the process of securing financial or nonfinancial return from value creation (Chesbrough et al., 2018). However, as we move value creation toward platform ecosystems, we need to address the system-centric perspective as well. Platforms present the highest potential for co-creation (Saebi & Foss, 2015), where an array of peripheral firms are connected to a central platform via shared or open-source technologies or technical standards with the objective to co-create value (Cennamo & Santaló, 2013; Jacobides et al., 2018). From the ecosystemic perspective, the logic is to enable value creation for all stakeholders, not only how it is captured by the focal firm (Upward & Jones, 2016; Zott et al., 2011). Thus, value co-creation may be identified as a *"joint, collaborative, concurrent, peer-like process of producing new value, both materially and symbolically"* (Galvagno & Dalli, 2014, p. 644).

The co-created value is co-captured through sharing and distributing the revenue between ecosystem members (Oh et al., 2015). Thus, value co-capture through platforms is realized through resource exploitation and coopetition, and through the provision of a unique platform by a focal actor for complementarities. However, commitment from the ecosystem actors and stakeholders invokes an extra dimension, value sharing, as potential stakeholders will not hand over their resources unless in return for what they can obtain from the relationship of the exchange (Verstraete & Jouison-Laffitte, 2011). Particularly, Adner (2022) stresses that for platforms to win at sharing value, it requires that both orchestrators and complementors work to further each other's interests.

Value networks and technology ecosystems of complementary actors have become central to successful value creation. With the emergence of these technology ecosystems, companies can combine capabilities across boundaries into innovative new service offerings and solutions to co-create and capture value (Aagaard, 2019). Thus, ecosystem partners face tensions from simultaneously sharing knowledge, while protecting internal knowledge to preserve their competitiveness (Rouyre & Fernandez, 2019). Therefore, to avoid the potential threat to the efficiency of value co-creation, the actors should consider simplifying the contracting and negotiation processes, formalizing knowledge transaction processes, and they need to implement transparent governance mechanisms (Zhu & Liu, 2018). These are typically handled by centralized rules and standards organized around a digital platform (Kretschmer et al., 2021). Thus, platform governance of value co-creation broadly concerns the design and deployment of governance choices, including decision rights, incentive structures, and control mechanisms. We, therefore, view ecosystem governance for value co-creation from a co-alignment structure perspective, which allows actors to productively combine their co-specialized inputs toward a joint value proposition (Adner, 2017; Thomas & Autio, 2020).

Value Creation in 5G

The service-centric definitions of standardized eMBB, URLLC, and mMTC do not characterize the currently available 5G services well. The first of the above, eMBB is under deployment globally, but the two latter industry-focused services URLLC and mMTC are yet to be adopted. Instead, we may find six types of commercially available mobile network constellation services: public commercial networks, public commercial virtual networks, neutral hosts, private local networks operated by MNOs, private local networks operated others than MNOs, and public network integrated non-public networks. Table 5.1 provides their description and key mechanisms of value co-creation and co-capture as well as their value sharing and spillover effects.

Table 5.1 Identified mobile communications services in 5G

	Description	Value co-creation & co-capture	Value sharing and spillover effects
Public commercial networks	Communications service provisioning for public use provided by MNOs	Anybody can subscribe to the service Connectivity services + other bundled services (equipment, content services)	Connectivity enables other players to provide their services to users. Net neutrality provides the basis for communications and the wide spillover effects throughout society
Public commercial virtual networks	Communications service provisioning for public use by a mobile virtual network operator (MVNO) utilizing an MNO's infrastructure	Anybody can subscribe to the service Connectivity services + other bundled services (equipment, devices, content services)	Opportunity for MNOs to monetize extra capacity Differentiation by user experience
Neutral host	Communications services for multiple MNOs' customers provided by an MNO or other than MNOs	Shared infrastructure costs and access to sites for MNOs Differentiation by quality of service challenged	Value sharing only between partners, spillover effects restricted

(continued)

Table 5.1 (continued)

	Description	Value co-creation & co-capture	Value sharing and spillover effects
Private local networks (NPN) operated by MNOs	Mobile network operator (MNO) offered local communications services for dedicated user group (private use)	Project and contract-based business between MNOs with specific customers, also internationally	Value sharing only between partners or a local ecosystem
Private local networks (NPN) operated by other than MNOs	Local stand-alone communications services for dedicated user group (private use) provided by other than MNOs	Value creation and capture by and for the local ecosystem without MNOs	Value sharing only between partners or a local ecosystem
Public network integrated non-public network (PNI-NPN)	Part of the network (e.g., slicing, access point name (APN) functionality) is shared between the local network owner and hosted by the MNO	Value creation focused on multi-local ecosystems that requires connectivity outside networks	Value sharing only between partners or a local ecosystem

Offering *public commercial networks* is the bread and butter of MNOs as these networks are open for all to subscribe to at the national level. The customer value comes primarily from enjoying the connectivity as a service, but often MNOs bundle connectivity with devices and various content services such as entertainment. Connectivity enables customers, whether consumers or organizations, to enjoy or provide various digital services to others. Public commercial networks are under net neutrality regulations and are obliged to provide their service so that all communications are treated equally without discrimination, offering users and other online content providers consistent rates irrespective of content, website, platform, application, type of equipment, source address, a destination address, or method of communications. Net neutrality regulations form the basis for the treatment of traffic for societal value creation and the high level of value spillover across the different sectors of society. *Public commercial virtual networks* are a category of public commercial networks that are operated by a different organizational entity using the infrastructure of another MNO. These networks are operated by mobile virtual network operators (MVNO)—that are often, but not always, MNO subsidiaries—that focus on specific customer segments requiring differentiated experience and a differentiated brand. For MNOs, an MVNO is an opportunity to monetize extra capacity. The MVNO spillover effects to other sectors or society mean increased competition and choice for end users.

A *neutral host* creates value by increasing cost efficiency by offering communications services for multiple MNOs serving their customers in given locations. Typically, all MNOs build their own mobile networks to provide the required coverage. In areas where the number of users is low, it may be that the returns on the network investment remain negative. In such areas, it might be beneficial for MNOs to collaborate and build only one network that serves all MNOs, thereby sharing infrastructure and site costs. This can be particularly effective in buildings where it is not feasible for all MNOs to build their own infrastructure. This neutral host concept can be run by one of the MNOs or some other organization so that all the MNOs' customers can utilize the service. Within the neutral host concept, value sharing primarily takes place between the partners, making the achieving of spillover effects restricted. The neutral host model can serve various campuses, factories, and hospitals, but also customers located in rural and sparsely populated areas.

Private local networks serve closed user groups and usually come in two variants. First, MNOs can offer local communications services for private dedicated user groups, starting with a project and continuing based on a specific contract. In this model, value is shared between the partners in the ecosystem, who are typically industrial customers. Second, private local networks can be offered by stakeholders other than MNOs, for example, local network companies, or end user organizations themselves can have local stand-alone communications services of their own. Additionally, in this case, the value creation and sharing takes place between the ecosystem partners, but now without MNOs. In cases where a connection outside the networks is needed, private local networks can also be integrated with public networks meaning that part of the network (e.g., slicing or access point functionality) is shared between the local network owner and hosted by an MNO. This kind of arrangement is called a public network integrated non-public network where value sharing takes place only between the ecosystem partners.

Strategic Considerations for 5G Services

From the value creation perspective, the key elements of mobile communications platforms include data, algorithms, components, and interfaces (Iivari et al, 2022; Yrjölä et al., 2021). *Components* include (readymade) add-on elements that connect to the platform to add functionality to it (Baldwin & Clark, 2000; Sanchez & Mahoney, 1996). *Interfaces*, such as specifications and design rules, describe how the platform and components interact and exchange information using well-documented and predefined standards such as application programming interfaces (APIs) (Katz & Shapiro, 1994). The engineering tradition has placed components and interfaces either at the core or periphery of the system. Baldwin (2008) found that modularity reduces coordination and transaction costs across the module boundary, while interface standardization reduces the asset specificity of modules. The increasing volume of *data* has transformed contemporary business practices (McAfee et al., 2012), while the *algorithm* revolution and cloud computing have given rise to a platform economy. Computing power is converted into economic tools using algorithms operating on data as the raw material.

In collaborative contexts, it makes sense to examine platforms regarding their openness from the combined perspectives of components, interfaces, data, and algorithms. Four strategies of openness can be identified for value creation. The traditional logic has been to base value creation on proprietary resources, but with increasing platformization, value creation strategies may be based on open interfaces (referred to as open edges of the platform) that the ecosystem stakeholders can utilize, open functionalities of the platform core that the stakeholders can access, or using an open-source approach where the stakeholders can themselves use and modify the platform's components, interfaces, and algorithms (Casadesus-Masanell & Llanes, 2011).

The six classes of mobile communications presented in Table 5.1 can be cross tabulated against the components, interfaces, data, and algorithms used in value creation and service delivery. At the same time, the openness of the system needs to be considered. The configuration of needs and resources in novel 5G digital services, therefore, implies a continuous balance between these different elements, where modularization can be considered as the basis for novel value configurations, as modularity can reduce coordination and transaction costs across the module boundary, while interface standardization can reduce the asset specificity of modules and extending the scalability of resources and services. Balancing user needs with resources implies a dualistic perspective on value, and both the *technical platform modularity and architecture* and *service modularity and architecture* (Yrjölä et al., 2021) need to be considered in the creation of new 5G services. The dualistic perspective on technical and service modularity and architecture can therefore be coined from the perspective of complementarity (Teece, 2018; Xu et al., 2020). Technology complementarity enables modularity from systems and encourages interoperability, whereas opportunity complementarity ensures a user-centric approach to service provisioning and service modularity. Complementarity relates to production, customers, asset prices, inputs, technologies, or innovation. As platforms with their ecosystemic characteristics can be either loosely or tightly coupled (Gawer, 2014; Teece, 2018), a focus on dynamic and integrative capabilities for value creation is important. Balancing between convergence, complementarity, and modularity is challenging, and therefore strategic thinking in terms of foresight and insight plays an influential role in ecosystems (Zahra & Nambisan, 2012).

6G-Enabled Value Creation

The value co-creation between companies in ecosystems is highly influenced by how technology evolves, as value and value propositions co-evolve along with the evolution of industry (Chau et al., 2020). With the technological evolution of 5G to 6G, new territories for value creation are being and will be established continuously. Recent research has explored some of the different typologies of value creation. For one, Kapoor and Teece (2021) discuss three different typologies of technological value creation: emerging, enabling, and embedding. As new technologies emerge (e.g., 5G and 6G), a trajectory is formed through a series of breakthrough inventions, which are introduced by a multiplicity of heterogenous ecosystem actors. This trajectory is associated with risks related to the emergent nature of the new technology. The enabling nature of technology corresponds to the commercialization of the technology across multiple application domains. This may be costly and require developments as well as an array of complementary assets, which may lead to underinvestment, hampering the growth and adoption of the technology. However, public policies and subsidies can support research and development activities of companies and ecosystems and may alleviate the situation. The embedded nature of technology captures the business model and ecosystem in which the technology is commercialized.

As 6G is still in the research phase, the use cases and services to be labeled as 6G-enabled are still emerging. At the same time, 6G is expected to become a future general-purpose technology with the potential to transform society. The new growth potential of 6G over earlier generations lies outside consumer services in 6G's capability to serve industries and smart cities, for example, thus boosting novel kinds of network effects and societal returns. First, for value creation, 6G services should meet the following *goals and expected impacts*: (1) to be human-centric and inclusive, (2) to be socially, environmentally, and economically sustainable (i.e., triple bottom line of sustainability), (3) to be trustworthy, and (4) to be resilient and maintain sovereignty. Second, 6G is expected to serve: (1) humans, (2) machines, (3) public and private organizations, and (4) communities as *users and developers*, suggesting the existence of different usage scenarios and examples as indications of 6G's role in everyday life and the different technological configurations and development trajectories expected to emerge for 6G in future. The new capabilities of 6G are expected to change the nature of human and machine life when communications merge with sensing and accurate positioning to serve humans and machines—and integrate the two—for increased efficiency, to enable earlier impossible tasks, and to automate processes. Finally, the question also remains of how to measure the performance of 6G with key performance indicators (KPIs) and value created with key value indicators (KVIs) (Hexa-X, 2021; Ahokangas et al., 2023).

The framework for IMT-2030 (6G) may cover four overarching aspects, which act as distinguishing design principles for 6G: sustainability, security/privacy/resilience, connecting the unconnected, and ubiquitous intelligence. No such design principles existed for 5G, making the design principles a novelty to be considered for 6G. 6G may thus embrace:

- Immersive Communication
- Hyper Reliable and Low-Latency Communications
- Massive Communication
- Ubiquitous Connectivity
- Integrated Artificial Intelligence and Communications
- Integrated Sensing and Communication

The following 6G-enabled services as sources of value creation can be identified:

- *Cost-efficient, sustainable, ubiquitous, near-instant, unlimited, mobile connectivity* as a basis for future connectivity-adjacent services for humans, organizations, and communities—and verticals such as industry, healthcare, logistics, or agriculture. The growing number of increasingly more autonomous things, also in swarms, can be considered as the users of 6G.
- *Multisensory applications and services* such as virtual, augmented, or extended mixed reality (VR, AR, and XR, respectively), holographic

communications and immersive telepresence to connect the physical, digital, and virtual worlds like the metaverse(s). For example, haptic and empathic communications may enable AI-enabled work in radically new ways.

- *Privacy, security, and safety* related services are increasing not only for humans in daily communications but also for ensuring that things, robots, and autonomous vehicles can be used safely, and that those critical infrastructures are secured. 6G may come with (also local) trust zones of different sizes and purposes.
- Massive *dynamic twinning*, i.e., the creation and existence of online and real-time digital twins (DT) of the physical reality for, e.g., smart factories or smart cities.
- *Transhumanism* via implanted biosensors and body-area networks to communicate and help merge humans and machines together, bringing humans new capabilities, creating a 'digital twin of me.'
- *Sustainable development* both at the societal and environmental level while being economically sustainable. First, trustworthy e-health services and institutional, local mobile coverage in schools and hospitals are examples of socially sustainable 6G services. Second, monitoring the earth via bio-friendly and energy-harvesting sensors exemplifies environmentally sustainable 6G.

The above list is not exhaustive, rather, it showcases many of the presently envisioned elements of 6G use. It can also be seen from different perspectives. Immersive communications, connecting intelligence, sensing for sustainability, and connecting the unconnected have been presented as 6G-specific usage scenarios or future examples (Ahokangas et al., 2023), exemplifying different sources, destinations, and types of value in future 6G. As an example, uncompromised triple bottom line sustainability is a new design principle for future mobile communications (Matinmikko-Blue et al., 2022). It is also evident that artificial intelligence will play a major role in future 6G value creation both as a stand-alone service but also as tightly integrated with 6G to enable novel services that enhance human and machine capabilities.

Trustworthy 6G

One of the recent challenges of 6G value creation concerns the question of trustworthy networks, specifically regarding privacy, security, safety, and resilience of mobile communications. On the one hand, achieving trustworthy mobile communications is a challenge for the whole lifecycle of network technologies from their definition to standardization, deployment, and use. For deployment and use, the EU Agency for Cybersecurity, ENISA (2022) lists ransomware, malware, social engineering, threats against data, threats against service and Internet availability, disinformation and misinformation, and supply chain attacks as potential threats that concern mobile communications. From the 6G system perspective, trustworthy networks require not only an understanding of the types and contents of threats as those listed by ENISA, but also the sources of these threats, whether individuals, organizations, or hostile nations. 6G needs to be designed and built to detect, prevent, and respond to these threats to ensure it remains trustworthy and resilient (Next G Alliance, 2022).

On the other hand, with the increasing use of artificial intelligence, 6G is expected to become more autonomous, which also increases the pressure on the trustworthiness of mobile communications. This can be seen also as an opportunity for future 6G. The 6G system characterized by architectural disaggregation, open interfaces, and embedded artificial intelligence, calls for trustworthy networks throughout the myriad of services and applications envisioned for the 6G ecosystem stakeholders (Ziegler et al., 2020). In the same way that artificial intelligence and the EU's AI act (EC, 2021) have done, 6G could benefit from being ethically responsible (Wu, 2022) with a focus on explainable transparency, fairness, accountability, robustness, safety, human agency and oversight, privacy, and data governance. However, trustworthy 6G calls for deep interaction between academia, industry, communities, and the authorities and regulators throughout its development. Amidst the existing geopolitical tensions, trustworthy 6G can also be considered a requirement for strategic autonomy, sovereignty, and democracy (Moerel & Timmers, 2021), which involves the increasing role of societal value, sustainability, and resilience in mobile communications.

Conclusions

This chapter has focused on value creation in 5G and future 6G by examining mobile communications services. The other side of the coin, value capture, will be discussed in subsequent chapters. It appears evident that platformization and ecosystems will play an increasingly important contextual role for mobile communications services as technologies converge and become integrated, making innovation efforts also more ecosystemic by nature. The original technology-driven value creation approach of the first mobile communications generations has evolved into a service-driven approach now that we have reached 5G. In light of extant research, it is possible that for 6G the value creation will become more human-centric and user and developer-driven, which in turn sets new demands on innovating applications and services for 6G. Furthermore, as value is continuously being created through multiple actors in ecosystems, platforms have taken over the 'process' of creating and capturing value. This development opens completely new venues for value co-creation as well as for sharing value between ecosystems. The managerial implications of these developments stress the necessity for strategic actions, while tailoring the companies' servitization efforts and platform strategies to leverage these unique business opportunities.

From the value creation perspective, 'ubiquitous wireless connectivity' in 6G will require that mobile operators integrate networks of different scales and scopes for the various needs of applications and users—from satellite networks to drone-based aerial, national terrestrial, down to various regional and local networks, even body-area networks, which all may be used for different purposes. This increasing variety of network technologies will create a multilayered multiplatform environment and will raise challenges in terms of developing services and identifying customers for value capture.

Bibliography

Aagaard, A. (2019). *Digital business models—Driving transformation and innovation.* Palgrave MacMillan.

Aagaard, A., & Ritzén, S. (2020). The critical aspects of co-creating and co-capturing sustainable value in service business models. *Creativity and Innovation Management, 29*(2), 292–302.

Aarikka-Stenroos, L., & Ritala, P. (2017). Network management in the era of ecosystems: Systematic review and management framework. *Industrial Marketing Management, 67*, 23–36.

Adner, R. (2017). Ecosystem as structure: An actionable construct for strategy. *Journal of Management, 43*(1), 39–58.

Adner, R. (2022). Winning platforms require that both leaders and followers work to further the other's interests. *MIT Sloan Management Review, 63*(1), 85–90.

Adner, R., & Kapoor, R. (2010). Value creation in innovation ecosystems: How the structure of technological interdependence affects firm performance in new technology generations. *Strategic Management Journal, 31*(3), 306–333.

Ahokangas, P., Matinmikko-Blue, M., & Yrjölä, S. (2023). Visioning for a future-proof global 6G from business, regulation and technology perspectives. *IEEE Communications Magazine, 61*(2), 72–78.

Amit, R., & Zott, C. (2012). Creating value through business model innovation. *MIT Sloan Management Review, 53*(2).

Appadurai, A. (1986). Introduction: Commodities and the politics of value. In A. Appadurai (Ed.), *The social life of things, commodities in cultural perspective* (pp. 3–30). Cambridge University Press.

Autio, E., & Thomas, L. D. W. (2019). Value co-creation in ecosystems: Insights and research promise from three disciplinary perspectives. In S. Nambisan, S. Lyytinen, K. Yoo (Eds.), *Handbook of digital innovation* (pp. 107–132). Edward Elgar, 107–132.

Baldwin, C. Y. (2008). Where do transactions come from? Modularity, transactions, and the boundaries of firms. *Industrial and Corporate Change, 17*(1), 155–195.

Baldwin, C. Y., & Clark, K. B. (2000). *Design rules: The power of modularity (Vol. 1).* MIT Press.

Barrett, M., Oborn, E., & Orlikowski, W. (2016). Creating value in online communities: The sociomaterial configuring of strategy, platform, and stakeholder engagement. *Information Systems Research, 27*(4), 704–723.

Bocken, N. M., Rana, P., & Short, S. W. (2015). Value mapping for sustainable business thinking. *Journal of Industrial and Production Engineering, 32*(1), 67–81.

Casadesus-Masanell, R., & Llanes, G. (2011). Mixed source. *Management Science, 57*(7), 1212–1230.

Cennamo, C., & Santaló, J. (2013). Platform competition: Strategic trade, tradeoffs in platform markets. *Strategic Management Journal, 34*(11), 1331–1350.

Chau, K. P., & Kan, C. W. (2020). The logic of innovative value proposition: A schema for characterizing and predicting business model evolution. *Journal of Business Research, 112*, 502–520.

Chesbrough, H., Lettl, C., & Ritter, T. (2018). Value creation and value capture in open innovation. *Journal of Product Innovation Management, 35*(6), 930–938.

Corsaro, D. (2020). Value co-destruction and its effects on value appropriation. *Journal of Marketing Management, 36*(1–2), 100–127.

Corvellec, H., & Hultman, J. (2014). Managing the politics of value propositions. *Marketing Theory, 4*(4), 355–375.

Cusumano, M. A., Yoffie, D. B., & Gawer, A. (2020). The future of platforms. *MIT Sloan Management Review, 61*(1), 26–34.

Davis, M. M., Spohrer, J. C., & Maglio, P. P. (2011). Guest editorial: How technology is changing the design and delivery of services. *Operations Management Research, 4*, 1–5.

EC. (2021). *Proposal for a regulation of the European parliament and of the council laying down harmonized rules on artificial intelligence (artificial intelligence act) and amending certain union legislative acts*. Regulation COM/2021/206 final.

Edvardsson, B., Frow, P., Jaakkola, E., Keiningham, T. L., Koskela-Huotari, K., Mele, C., & Tombs, A. (2018). Examining how context change foster service innovation. *Journal of Service Management, 29*(5), 932–955.

El Sawy, O. A., & Pereira, F. (2013). *Business modelling in the dynamic digital space: An ecosystem approach*. Springer.

ENISA. (2022). *ENISA threat landscape*. https://doi.org/www.enisa.europa.eu/publications/enisa-threat-landscape-2022/@@download/fullReport.

Galvagno, M., & Dalli, D. (2014). Theory of value co-creation: A systematic literature review. *Managing Service Quality, 24*(6), 643–683.

Garcia-Swartz, D. D., & Garcia-Vicente, F. (2015). Network effects on the iPhone platform: An empirical examination. *Telecommunications Policy, 39*, 877–895.

Gawer, A. (2014). Bridging differing perspectives on technological platforms: Toward an integrative framework. *Research Policy, 43*(7), 1239–1249.

Geissdoerfer, M., Vladimirova, D., & Evans, S. (2018). Sustainable business model innovation: A review. *Journal of Cleaner Production, 198*, 401–416.

Grover, V., & Kohli, R. (2012). Cocreating IT value: New capabilities and metrics for multifirm environments. *MIS Quarterly*, 225–232.

Hexa-X Project. (2021). *Deliverable D1.2. Expanded 6G vision, use cases and societal values.* https://hexa-x.eu/wp-content/uploads/2022/04/Hexa-X_D1.2_Edited.pdf.

Hofacker, C. F., & Corsaro, D. (2020). Dystopia and utopia in digital services. *Journal of Marketing Management, 36*(5–6), 412–419.

Iivari, M., Ahokangas, P., Matinmikko-Blue, M., & Yrjölä, S. (2022). Opening closed business ecosystem boundaries with digital platforms: Empirical case of a port. In *Emerging ecosystem-centric business models for sustainable value creation* (pp. 67–96). IGI Global.

ITU-R. (2015). IMT Vision–Framework and overall objectives of the future development of IMT for 2020 and beyond. *Recommendation ITU, 2083*(0). https://www.itu.int/rec/R-REC-M.2083.

Jacobides, M. G., Cennamo, C., & Gawer, A. (2018). Towards a theory of ecosystems. *Strategic Management Journal, 39*(8), 2255–2276.

Kallinikos, J. (2011). Bureaucracy under siege: On information, collaboration and networks. In S. R. Clegg, M. Harris, & H. Höpfl (Eds.), *Managing modernity: Beyond bureaucracy?* (pp. 132–154). Oxford University Press.

Kapoor, R., & Teece, D. J. (2021). Three faces of technology's value creation: Emerging, enabling, embedding. *Strategy Science, 6*(1), 1–4.

Katz, M. L., & Shapiro, C. (1994). Systems competition and network effects. *Journal of Economic Perspectives, 8*(2), 93–115.

Kretschmer, T., Leiponen, A., Schilling, M., & Vasudeva, G. (2021). Platform ecosystems as meta-organizations: Implications for platform strategies. *Strategic Management Journal, 3*(3), 405–424.

Lepak, D. P., Smith, K. G., & Taylor, M. S. (2007). Value creation and value capture: A multilevel perspective. *Academy of Management Review, 32*, 180–194.

Matinmikko-Blue, M., Yrjölä, S. & Ahokangas, P. (2022). Sustainable technology design for future 6G mobile communications. In L. Filho (Ed.), *Handbook of sustainability science in the future*. Springer.

McAfee, A., Brynjolfsson, E., Davenport, T. H., Patil, D. J., & Barton, D. (2012). Big data: The management revolution. *Harvard Business Review, 90*(10), 60–68.

McIntyre, D. P., & Srinivasan, A. (2017). Networks, platforms, and strategy: Emerging views and next steps. *Strategic Management Journal, 38*(1), 141–160.

Moerel, L., & Timmers, P. (2021). *Reflections on digital sovereignty.* EU Cyber Direct, Research in Focus series.

Next G Alliance. (2022). *Trust, security, and resilience for 6G systems.* https://www.nextgalliance.org/white_papers/trust-security-and-resilience-for-6g-systems/.

Oh, J., Koh, B., & Raghunathan, S. (2015). Value appropriation between the platform provider and app developers in mobile platform mediated networks. *Journal of Information Technology, 30*(3), 245–259.

Orlikowski, W. J., & Scott, S. V. (2008). Sociomateriality: Challenging the separation of technology, work and organization. *Academy of Management Annals, 2*(1), 433–474.

Parida, V., Sjödin, D., & Reim, W. (2019). Reviewing literature on digitalization, business model innovation, and sustainable industry: Past achievements and future promises. *Sustainability, 11*(2), 391.

Patala, S., Jalkala, A., Keränen, J., Väisänen, S., Tuominen, V., & Soukka, R. (2016). Sustainable value propositions: Framework and implications for technology suppliers. *Industrial Marketing Management, 59*, 144–156.

Pedersen, E. R. G., Gwozdz, W., & Hvass, K. K. (2018). Exploring the relationship between business model innovation, corporate sustainability, and organisational values within the fashion industry. *Journal of Business Ethics, 149*(2), 267–284.

Phillips, M. A., & Ritala, P. (2019). A complex adaptive systems agenda for ecosystem research methodology. *Technology Forecasting & Social Change, 148*, 119739.

Prahalad, C., & Ramaswamy, V. (2004). *The future of competition: Co-creating unique value with customers.* HBS Press.

Rai, A., Constantinides, P., & Sarker, S. (2019). Next generation digital platforms: Toward human-AI hybrids. *MIS Quarterly, 43*(1), iii–ix.

Rouyre, A., & Fernandez, A. (2019). Managing knowledge sharing-protecting tensions in coupled innovation projects among several competitors. *California Management Review, 62*(1), 95–120.

Ruutu, S., Casey, T., & Kotovirta, V. (2017). Development and competition of digital service platforms: A system dynamics approach. *Technological Forecasting & Social Change, 117*, 119–130.

Saebi, T., & Foss, N. J. (2015). Business models for open innovation: Matching heterogeneous open innovation strategies with business model dimensions. *European Management Journal, 33*(3), 201–213.

Sanchez, R., & Mahoney, J. T. (1996). Modularity, flexibility, and knowledge management in product and organization design. *Strategic Management Journal, 17*(2), 63–76.

Shipilov, A., & Gawer, A. (2020). Integrating research on inter-organizational networks and ecosystems. *Academy of Management Annals, 14*, 92–121.

Sorri, K., Seppänen, M., Still, K., & Valkokari, K. (2019). Business model innovation with platform canvas. *Journal of Business Models, 7*(2), 1–13.

Stark, D. (2009). *The sense of dissonance: Accounts of worth in economic life.* Princeton University Press.

Teece, D. J. (2018). Profiting from innovation in the digital economy: Enabling technologies, standards, and licensing models in the wireless world. *Research Policy, 47*(8), 1367–1387.

Thomas, L. D. W., & Autio, E. (2020). *Innovation ecosystems in management: An organizing typology.* Oxford University Press.

Upward, A., & Jones, P. (2016). An ontology for strongly sustainable business models: Defining an enterprise framework compatible with natural and social science. *Organization and Environment, 29*, 97–123.

Vargo, S. L., Lusch, R. F., & Akaka, M. A. (2010). Advancing service science with service- dominant logic. In P. P. Maglio, C. A. Kieliszewski, & J. Spohrer (Eds.), *Handbook of service science* (pp. 133–156). Springer.

Vasudeva, G., Leiponen, A., & Jones, S. L. (2020). Dear enemy: The dynamics of conflict and cooperation in open innovation ecosystems. *Strategic Management Review, 1*, 355–379.

Verstraete, T., & Jouison-Laffitte, E. (2011). *A business model for entrepreneurship.* Edward Elgar Publishing.

Wu, Y. (2022). Ethically responsible and trustworthy autonomous systems for 6G. *IEEE Network, 36*(4), 126–133.

Xu, Y., Kemppainen, L., Ahokangas, P., & Pikkarainen, M. (2020). Opportunity complementarity in data-driven business models. *Journal of Business Models, 8*(2), 92–100.

Ye, H., & Kankanhalli, A. (2018). User service innovation on mobile phone platforms: Investigating impacts of lead userness, toolkit support, and design autonomy. *MIS Quarterly, 42*(1), 165–187.

Yrjölä, S., Ahokangas, P., & Matinmikko-Blue, M. (2021). Novel platform-based ecosystemic business models in the future mobile operator business. *Journal of Business Models, 9*(4), 67–93.

Zahra, S. A., & Nambisan, S. (2012). Entrepreneurship and strategic thinking in business ecosystems. *Business Horizons, 55*(3), 219–229.

Zhu, F., & Iansiti, M. (2019). Why some platforms thrive and others don't. *Harvard Business Review,* 118–125.

Zhu, F., & Liu, Q. (2018). Competing with complementors: An empirical look at Amazon.com. *Strategic Management Journal, 39*(10), 2618–2642.

Ziegler, V., Viswanathan, H., Flinck, H., Hoffmann, M., Räisänen, V., & Hätönen, K. (2020). 6G architecture to connect the worlds. *IEEE Access, 8,* 173508–173520.

Zott, C., Amit, R., & Massa, L. (2011). The business model: Recent developments and future research. *Journal of Management, 37*(4), 1019–1042.

6

Business Models in 5G/6G Mobile Communications

Petri Ahokangas, Annabeth Aagaard, Irina Atkova,
Seppo Yrjölä, and Marja Matinmikko-Blue

A process cannot be understood by stopping it. Understanding must
move with the flow of the process, must join it and flow with it.

Frank Herbert, Dune

P. Ahokangas (✉) · I. Atkova
Martti Ahtisaari Institute, Oulu Business School, University of Oulu, Oulu,
Finland
e-mail: petri.ahokangas@oulu.fi

I. Atkova
e-mail: irina.atkova@oulu.fi

A. Aagaard
Department of Management, Aarhus University, Aarhus, Denmark
e-mail: aaa@mgmt.au.dk

S. Yrjölä
Centre For Wireless Communications, University of Oulu, Oulu, Finland

© The Author(s) 2024   **137**
P. Ahokangas and A. Aagaard (eds.), *The Changing World of Mobile Communications*,
https://doi.org/10.1007/978-3-031-33191-6_6

The Importance of Business Models in Mobile Communications

Mobile networks have become the backbone for the digitalization of society, making mobile network operators (MNOs) one of the key players of the modern digitalized society (Li & Whalley, 2002). One of the modern tools for making sense of and communicating digitalization is the business model (Timmers, 1998), which explains how a business creates and captures value (Amit & Zott, 2001) as a *process*. For MNOs, the traditional business model has been to monetize mobile connectivity for consumer and corporate end users—bundled with dealership of digital content and/or equipment, also installed—and differentiated by the quality of service, coverage, or data rates/quotas, based on exclusive use of spectrum (Ahokangas et al., 2021a). The business models employed by MNOs to offer ubiquitous mobile connectivity radiate their impact on all current digital services. Without connectivity, no digital content could be sent or received. Without the abundance of content, digital context services such as search engines or combined data, user, and location information would be of low value; and commerce platforms would lack merchandise. Additionally, without connectivity, the value of artificial intelligence cannot be realized. However, the above-described primary business models of the mobile network operators will be disrupted by the fifth generation of mobile communications (5G) currently being introduced. One example of this disruption is the emergence of the local (or micro) operator concept that complements the

Nokia, Oulu, Finland

S. Yrjölä
e-mail: seppo.yrjola@oulu.fi; seppo.yrjola@nokia.com

M. Matinmikko-Blue
Infotech Oulu Focus Institute and Centre for Wireless Communications, University of Oulu, Oulu, Finland
e-mail: marja.matinmikko@oulu.fi

traditional nation-wide MNO services through local and often private networks for tailored use (Matinmikko et al., 2017).

Adding to the enhanced mobile broadband of the present 5G technology, the increasing softwarization and cloudification of 5G networks will help in future to serve the varying needs of new types of users such as machines, autonomous vehicles, drones, robots, and communities in critical and massive machine-to-machine communications, also using shared spectrum. With a service-centric approach, 5G was originally defined through three technical usage scenarios: enhanced mobile broadband (eMBB), ultra-reliable low-latency communications (URLLC), and massive machine-type communications (mMTC) (ITU-R, 2015). With higher frequencies and higher bandwidth, 5G means smaller cell sizes that enable local and private 5G networks for different verticals that have specific requirements (Ahokangas et al., 2021b), also indoors. Consequently, it has been argued that the whole MNO-centric ecosystem, its stakeholders, and the business models therein will change in future 5G (Matinmikko et al., 2018), giving the floor to a variety of new operator concepts.

Indeed, the term, 'telecommunications service provider,' as used for mobile network operators providing telecommunications services, is subject to specific regulatory rights and obligations (Matinmikko et al., 2017) that might not exist in all cases of local networks and may vary between countries. Consequently, in this chapter, we use the generic term *mobile operator* when discussing future business models. These disruptive changes call for exploring and understanding what 5G and later 6G will mean in the mobile communications business model context and what the implications are for the business model content, structure, and governance (Amit & Zott, 2001).

Strategy and Technology at the Core of the Business Model

The phenomenon of a business model has conceptually matured by drawing insights, among others, from the field of strategy and technology. Practically, technological development and the subsequent emergence of e-commerce in the mid-1990s brought the term business model into the vocabulary of managers and scholars. Back in the day, the term was actively used in electronic markets to describe and explain how value could be captured by buying and selling products and services over the electronic network. One of the first definitions of the business model concept developed in the technological field is the iconic definition by Timmers (1998, p. 4) who explains a business model as "*an architecture for the product, service and information flows including a description of the various business actors and their roles, the potential benefits for the various business actors, and the sources of revenues.*" Over the years, scholars have debated the definition of the business model concept and nowadays increasingly converge on the idea that value creation, delivery, and capture mechanisms constitute the backbone of the concept (Ritter & Lettl, 2018). Further, proliferation of the Internet and subsequent emergence of the new competitive market structures have created fertile ground for the formation of a magnitude of various value-related mechanisms.

To systematize the growing palette of the digital activities, Wirtz et al. (2010) developed the 4C typology of Internet business models. In the content business model, value creation, delivery, and capture mechanisms are organized to provide users access to various types of digital content. The commerce business model can be viewed as a predecessor of a platform business model, in which the main value proposition is to provide an exchange place for buyers and sellers. Context-oriented business models focus on aggregating information for the users to ensure seamless navigation and reduce the complexity and non-transparency of the digital environment. Connection-oriented business models, as the name suggests, are aimed at providing physical or virtual network infrastructure. In this, connectivity enables stronger inter-firm collaboration and supports the development of digitally enabled ecosystems (Miehe et al., 2022).

The Concept of the Business Model

Theoretically, the concept of a business model is deeply rooted in the strategic management field and therefore, the evolution of the business model definition reflects the increasing importance of the strategic components of business models (Morris et al., 2005). Strategic management research enriched the business model discourse with the concepts of opportunity, value, and advantage, wherein a business model serves as a vehicle for a coherent implementation of strategy (Dahan et al., 2010). For instance, Morris et al., (2005, p. 727) define a business model as "*a concise representation of how an interrelated set of decision variables in the areas of venture strategy, architecture, and economics are addressed to create sustainable competitive advantage in defined markets.*" A strategy and technology orientation are fundamental in the business model research field and several attempts have been made to bridge the divide. Chesbrough and Rosenbloom (2002, p. 529) explain the concept of a business model as "*the heuristic logic that connects technical potential with the realization of economic value.*" The phenomenon of platformization and the emergence of platform business models have further integrated the strategic and technological roots of the business model concept (Nambisan et al., 2019).

Overall, the business model has become a contemporary paradigm for exploring and exploiting different business-related ideas and conceptualizations (Wirtz et al., 2016). Even in the absence of a commonly accepted definition, the extant literature depicts the business model as a boundary-spanning, multi-purpose, and futures-oriented vehicle for designing, doing, and making sense of digital business (Zott et al., 2011). For example, Massa et al. (2017) see business models as addressing how firms do business, how this is interpreted, or how a business model could be represented through formal conceptualizations. However, regardless of the lack of a common definition, there are an abundance of business model templates and tools that can be used to describe and design business models. The business model scholars appear to be unanimous that the primary function of a business model is to explore and exploit a business opportunity. In turn, the opportunity sets the logic for the organization of the value-related processes. Together, the opportunity and value processes set the stage for formulating competitive advantage (Chesbrough, 2010; Zott & Amit, 2010). In turn, the sustainability of competitive advantage is contingent upon its replicability (Chaharbaghi & Lynch, 1999). Additionally, digitalization and proliferation of the ecosystemic approach in the business model literature have brought business model scalability into the discussion (Nielsen & Lund, 2018).

Regulation, Technology, and Business

The traditional way to look at businesses in mobile communications has been to explore the changes in the regulative and technological domains, both having a significant impact on business decisions, especially the business models employed by the operators (Ahokangas et al., 2013). Spectrum and competition regulations have played a pivotal role regarding the business models applied by operators, either allowing, delimiting, or protecting/safeguarding certain business models. Technology, in turn, has been the business model enabler and a driver for competitive edge and competition with new and improved services, while also 'pushing' the operators to innovate and diversify their offerings. However, up to the fourth generation of mobile communications (4G) networks, the primary business models applied by leading operators have remained surprisingly unchanged (Lehr et al., 2021), although they have been seriously challenged by the content-owning, cloud-based over-the-top (OTT) Internet giants. Being challenged by the OTTs, many operators' margins and revenue have started to deteriorate. As operators are struggling with whether and how to innovate their business models in practice, the question arises as to what kind of an approach would be appropriate to understand future operators' business models in 5G and 6G and what these novel business models would look be. As new forms of operators are expected to emerge in future (Matinmikko et al., 2017), it is crucial to map the factors according to which the emergence of these operators and their respective business models could be outlined.

Internationalization of Business Models in Mobile Communications

Related to business models, an astonishingly little researched topic in mobile communications is the internationalization of mobile operators. The direction and extent of mobile operators are defined by national regulations and policies, in Europe also by the EU-level regulations, which have a direct impact on the business models used in the industry. Although being a global business, the mobile communications business is highly regulated at the national level. The internationalization of the industry started in developed countries with the liberalization of markets in the 1990s when traditional state monopolies were transformed into business entities. The period 1990–2010 could be characterized as the era of emergence and rise of MNOs and seen as a period of rapid internationalization of connectivity (Gooderham et al., 2022). During 2010–2020, the OTTs overran MNOs with their content-based business models. After 2020, Gooderham et al. (2022) envisioned MNOs to face marginalization unless they paid serious attention to their business models.

Dike and Rose (2017) carried out a systematic analysis of the internationalization of mobile telecommunications and summarized the key motivations for internationalization in the sector:

- Business-friendly regulatory regimes in potential host countries.
- Increased competitive pressure in home countries associated with decreasing domestic growth potential.
- Increasing subscriber acquisition costs; and shrinking average revenue per user (ARPU).
- Previous internationalization experience.

- Strategic factors related to scale and scope advantages of business.

In Europe, major MNOs such as Deutsche Telekom (Germany), Telefonica (Spain), Orange (France), and Vodafone (UK) are examples of highly international operators with a presence mostly through mergers and acquisitions in several European countries but also in Africa, South and North America, and Africa. Other MNOs such as Telia (Sweden) have also entered East European and former Soviet Union markets. Within Africa, the internationalization patterns of African operators have not followed traditional internationalization theory in terms of location choices (i.e., not prioritized neighboring countries) or country characteristics (i.e., selected countries with the highest growth potential; see Dike & Rose, 2019). African companies have also tended to adjust their strategies to local conditions rather than trying to leverage their firm-specific competencies (Jahanbakht et al., 2022).

For years, the Non-Terrestrial Network (NTN) satellite communication has remained standalone global technology, independent of national mobile terrestrial communication networking. Recently, the R&D interest in NTNs in academia and industry has increased (Rinaldi et al., 2020), and commercial solutions are emerging with worldwide deployments associated with internationally applied business models. With the next generation of satellites, initially based on 5G architecture, NTN will integrate with terrestrial networks with the main objective to provide ubiquitous global coverage to user devices for consumers and industries, particularly in unserved and underserved areas. The NTN component is envisioned to become essential within the 6G ecosystem to ensure service availability, continuity, ubiquity, and scalability.

An emerging new model for internationalization is currently taking place in the context of local and private networks. The real challenge for local networks is their international scalability and replicability as many of these networks require considerable tailoring and in-depth understanding of customers' needs. Therefore, integrators and cloud companies that are specializing in specific industries and their needs—and with the capabilities to plan, install, and maintain local networks—and that have an extensive international presence and local partnerships, are building efficient replication-based internationalization strategies in this fast-growing new niche.

Building on earlier research, we see operators' business models accumulating value on platforms and ecosystems as enabled by technology and delimited by regulation. From a business model perspective, we consider the future mobile communications system as a dynamic connectivity platform converging with various (other) digital platforms, thus forming a platform ecosystem comprised of complementary business models that are not necessarily hierarchically controlled by any of the stakeholders of the emerging ecosystem. As superior business models can help successfully commercialize mediocre technologies (Chesbrough, 2010), technology can be considered as an antecedent to the business model. From these starting points, this chapter aims to contribute by analyzing current and future mobile operator business models.

Business Models for 5G Mobile Operators

In the mobile communications context, Al-Debei and Avison (2010) present a business framework comprising the dimensions of the value proposition, architecture, network, and finance. As one of the early works on this topic, the paper followed the traditional business model approach of the time. The classification of connectivity, content, context, and commerce business models (4C) made for the internet 2.0 business models (Wirtz et al., 2010) helps to characterize mobile communications businesses. Within mobile communications, the 4C typology of business models can be interpreted as nested layers, where the lower layer business models of connectivity and content are required as enablers and value levers for the higher layers of the business models, the context and commerce, to exist. Traditionally, MNOs have offered connectivity in a mass-produced mode, with price, data rates, quotas, or coverage as differentiation (Ahokangas et al., 2021a). Additionally, some operators have started to offer bundled content—such as entertainment—or equipment as a dealer. Personalized or tailored services such as context (i.e., location-based, service-specific, or data-based) or commerce (i.e., platform-enabled ubiquitous services) business models have often been separated from the connectivity business. The only exception to tailored services have been big enough industrial customers with vertical-specific needs, and these have often been served in collaboration with network equipment vendors, network constructors, and service integrators. There also exist operators that specialize in servicing industrial customers and their IoT (internet-of-things) needs.

Mobile networks can be regarded as connectivity platforms or ecosystems, depending on the perspective. Technically a platform can be divided into a centralized core and geographically distributed access networks. The core network takes care of the services and billing, while the access networks—which can currently comprise several technology generations from 2G up to 5G—provide the radio access from a variety of user devices to the networks. With 5G, mobile platforms are increasingly becoming combined or converged with various digital platforms of cloud services and OTT internet service providers, while enabling platform ecosystems (Gawer & Cusumano, 2014) or the sharing economy (Ahokangas et al., 2021a). This relationship between business models builds on platforms, and several researchers have addressed the networked or ecosystemic nature of the business environment.

A Technology-Dominated View of Business Models

A review of earlier research on MNOs' business models reveals the fundamental technical starting points of the extant research (Yrjölä et al., 2022). A widely used business model approach within mobile communications is the 'as-a-service' logic (Ives & Learnmonth, 1984) that can be divided into scalable infrastructure-as-a-service (IaaS), platform-as-a-service (PaaS), software-as-a-service (SaaS), and data-as-a-service (DaaS) up to everything/anything-as-a-service (XaaS) models (Duan et al., 2015) with the extensive use of algorithms. In this technical line of research, Noll and Chowdhury (2011) introduced technology-enabled collaborative business models, while Rasheed et al. (2015) presented the brokerage business models, and Zhang et al. (2015) discussed a cloud-assisted model. Beyond technicalities, these all represent two primary mobile operator business models, that of a *connectivity service provider* and its *differentiation toward content services.*

Rao and Prasad (2016) identified the mobile broadband (MBB) business model, the target expansion business model with a focus on other than consumer customers, and the outsourced managed services business model, where the network infrastructure providers offer the network as a service (NaaS). Rao and Prasad (2016) also identified the mobile virtual network operator (MVNO) business model, where a separate entity, often a mobile network operator's subsidiary, offers segmented services by using the infrastructure of a 'real' operator. Furthermore, they predicted the evolution of business models toward digital business models in the forms of various connectivity providers and partnership business models. Lindgren (2016) discussed persuasive business models by paying attention to their physical, digital, and virtual dimensions.

Camps-Aragó et al. (2019) examined MNOs' business models. They presented a classification to a micro-operator, the cloud-based XaaS/NaaS, the use case enabler for business-to-business customers, the ecosystem orchestrator, and the pervasive platforms business model. Kukliński et al (2018) discussed business models for network slicing, proposing technical role-based business models for infrastructure brokers, network slice brokers, and service brokers. Hmoud et al. (2020) discussed mobile network operator business models targeted for two-sided markets and presented big data-driven (i.e., based on crowdsourced data), advertising application (i.e., based on advertising platform), and mobile sensing (i.e., monitoring users' equipment for location or activity) based business models. Finally, Sacoto-Cabrera et al. (2020) analyzed the monopolistic and strategic business models of mobile network and mobile virtual network operators using game-theoretic modeling, concluding that both business models were economically sustainable.

The Strategic Approach to 5G Mobile Network Operator Business Models

Another stream of literature on business models has adopted a more strategy-oriented approach, classifying mobile network operators

(MNOs) based on their scale and scope and looking at their scalability, replicability, and sustainability. Matinmikko et al. (2017) proposed that local 5G micro-operators could run bundled connectivity (i.e., local connectivity), content (e.g., locally tailored services), context (e.g., secure local networks for vertical-specific needs), or commerce (e.g., 'my data' operator services) business models. Ahokangas et al. (2021a) identified two types of future mobile network operator BMs: the general bit-pipe and segmented specialized service business models, thus drawing a line between connectivity- and content-based BMs. In addition, the authors identified the wholesale service, retail service, context service, and vertical service business models for local operators. The resulting 5G mobile operator business models can be presented as follows:

- *The General Bit-Pipe MNO* business model is a future projection of today's dominant nation-wide MNO model with a large installed base who utilize a variety of mobile communication technologies from 2G to 5G, often complemented by Wi-Fi and IoT technologies to provide general mobile broadband connectivity to all in a mass-production bit-pipe mode. This business model can be used to offer public commercial networks or public commercial virtual network services.
- *The Segmented Specialized Service MNO* business model builds on offering mobile connectivity bundled with specialized content to selected segments nation-wide or regionally. These operators are the challengers to dominant MNOs and have a smaller installed base and attempt to compete where the general bit-pipe operators are less competitive serving the long tail of customers through higher value-added services. This business model can be used to offer services such as public commercial networks, public commercial virtual networks, neutral hosts, or private local networks.
- *The Wholesale Service Local Operator* business model builds on the opportunity to offer local hosted connectivity to MNOs' customers as a neutral host. This is an opportunity in public, but restricted places, such as campuses and hospitals where it is not feasible that all MNOs build their own network infrastructure but outsource it from a local operator. The local operator would then directly charge the MNOs, not the end users, for the service.

- *The Retail Service Local Operator* business model is based on offering local connectivity and complementary data services to end users in venues such as shopping malls, hotels, or sharing workplaces/offices independent of MNOs. This business model may serve MNOs to provide private local networks or public network integrated non-public networks.
- *The Vertical Service Local Operator* business model is about offering private local networks, i.e., connectivity, content, and context services for verticals like factories, campuses, and ports independent of MNOs. The users of the service could be either humans or machines.
- *The Context Service Local Operator* business model builds on offering personalized consumer services (via private local networks) or networks including connectivity and content and context data on-demand using, e.g., network slicing technology.

The above business model examples are indicative of a new kind of business ecosystem that is expected to emerge around 5G including not only the MNOs and their users, but also mobile network infrastructure vendors, facility owners and tenants, network infrastructure constructors, data and other content providers, and end user and other equipment manufacturers will be able to adopt new business roles. This ecosystem can be vertically, horizontally, or obliquely structured. In vertically struc-tured ecosystems, value accumulates from the suppliers toward the demand-side customers, conceptually separating value creation, delivery, and capture. In horizontal structures, value is co-created and co-captured in stakeholder interaction in two-sided markets. Oblique ecosystem structures indicate the emergence of a multisided platform ecosystem, where value co-creation and co-capture can take place through multiple roles in the ecosystem with value spillovers to upstream and downstream players (Iivari et al., 2016). In the above categorization, the four first can be labeled as horizontal models, the fifth a vertical model, and the last one an oblique model.

A Strategy-Technology View

Up to now, most of the above-presented 5G-enabled business models call for future development and deployment. However, the discussion gives rise to a framework that depicts (Fig. 6.1) the business model discussions in the mobile communications context from the strategy and technology viewpoints. The strategy viewpoint comprises two perspectives: the traditional opportunity and value creation elements of business models, which highlight the role of novel advantages required for mobile communications businesses, and the traditional connectivity element that is complemented by the novel content, context, and commerce elements of mobile communications businesses. The technology viewpoint comprises the need to consider the scalability and replicability of business models, giving rise to the increased importance of sustainability aspects in future business models. The technology viewpoint emphasizes the increasing role of technology in supporting, fostering, and driving the scalability and replicability of business models. The greater scalability and replicability of the business models help to incorporate and further realize sustainability goals such as energy savings and decreased CO_2 emissions/environmental pollution. In addition, the technology viewpoint considers the platform perspective that traditionally covers components and interfaces to also include data and algorithms (Yrjölä et al., 2021). Overall, it can be concluded that mobile communications business models are not easy to depict in simple terms. Making sense of the technology-oriented business model literature requires an understanding of the technological concepts—like cloud stacks and platforms or network slicing—used in the discussion.

Envisioned Future 6G Business Models

The modern 5G business model context can increasingly be characterized as a VUCA environment: volatile, uncertain, complex, and ambiguous (Bennett & Lemoine, 2014). This implies that to deal with the VUCA challenges, it has become increasingly crucial to deal with the dynamism of business models in their respective changing business environments.

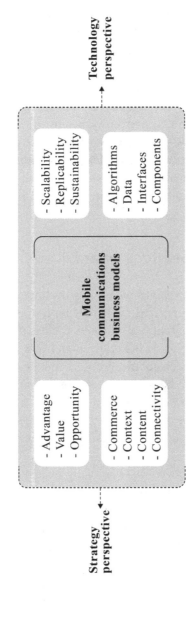

Fig. 6.1   A strategy-technology view on mobile communications business models

Business models always function in a context and therefore need to be calibrated to their respective business context (Teece, 2010). For future 6G, this means that business model innovation needs to be extended from the business model level to the ecosystem level (Snihur & Bocken, 2022), emphasizing sustainability. The first 6G white paper on 6G (Yrjölä et al., 2020) envisioned interactive business model configurations based on differing *needs and demands* of humans, machines, organizations, and communities and the various *assets, resources, and capabilities* residing in the ecosystem, thereby identifying *matching, bridging, brokering,* and *sharing*-based business models for the future 6G ecosystem(s). Generally, it can be considered that digitalization is driving toward converging multi-platform ecosystems, where business models may be reconfigurable, and firms may run several business models in parallel.

Business Model Innovation Toward 6G

The compound effects of various technology enablers, emerging regulatory delimitations, and integrated triple bottom line economic, environmental, and social sustainability on business models call for a discussion of business model innovation in 6G. Currently, 6G is still in the research phase. However, 6G has been envisioned as a general-purpose technology platform or infrastructure that necessitates ecosystemic innovation, as no single firm can alone develop it. Up to now, the telecommunications industry has followed the *define-standardize/implement-deploy/use* cycle of technology generation commercialization based on standard releases (Ahokangas et al., 2023a).

For business model innovation—especially related to finding scalable business opportunities—this implies new societal and environmental requirements, regulations, and stakeholders to be considered at each of the stages and releases of technology. For sustainable value creation, the diverging field of standardization and new integrated technologies with diverse development trajectories and competing implementations set increasing pressures for foresight-based strategies for technology deployment and use. Further, for replicating the technology-based competitive advantages in different markets or customer segments, the ubiquitous mobility of 6G sets demands for novel kinds of collaboration.

Already today, regulation and sustainability go hand in hand influencing mobile operators' business models in two ways. First, policymakers are concerned about the energy efficiency of mobile networks. In times of increasing energy costs, the need to make 5G and 6G more energy efficient is an economic motivator for mobile operators to save costs, especially operating expenses (OPEX), but also to reduce CO_2 emissions. Although the ICT industries have been so far excluded from CO_2 compensation requirements, it could be considered that in future this may change. Further, there are increasing concerns regarding the electromagnetic fields (EMF) caused by mobile communications and the consumption of critical and rare raw materials, that indirectly and directly set demands and limitations on business model innovation. Additionally, the critical role of mobile communications sets demands on developing and maintaining the security and resilience of networks to ensure societal sustainability. Trustworthiness via security considerations needs to cover all the aspects of cybersecurity, including resilience against attacks, preservation of privacy, and ethical, safe application of automation to network operations and applications. For the same reason, regulations related to strategic autonomy and sovereignty have been introduced in many countries.

Envisioned Business Models in 6G

Research on 6G business models is yet scarce. The present literature mostly emphasizes 5G and beyond business models from a technology perspective (Yrjölä et al., 2022). Following the ITU-R use cases presented for 5G, one starting point for 6G comes from the European Hexa-X project that has identified five use case families for 6G: sustainable development for both environmental and social sustainability; massive digital twinning of physical reality; immersive telepresence in human-to-human communications; from robots to cobots (enabling collaborative robots); and local *trust zones* for trustworthy communications between humans and machines (Hexa-X, 2021). Following the trends of converging platforms and the increasing importance of data and artificial intelligence-driven digitalization, four novel 6G use cases or service categories have been presented: connecting the unconnected, connecting intelligence, sensing for sustainability, and immersive communications (Ahokangas et al., 2023b).

The different versions of the future metaverse—consumer, enterprise, or industry—can be seen at the core of future 6G solutions. However, it can be expected that the metaverse will spread to all areas of human life. Thus, from technology, business, and regulation perspectives, the metaverse may emerge through different trajectories and be applied in various domains. From a technology perspective, the metaverse needs integrated 6G and AI as the basis for immersive communications. Barrera and Shaf (2023, p. 6) defined the metaverse as: "*the technology-mediated network of scalable and potentially interoperable extended reality environments merging the physical and virtual realities to provide experiences characterized by their level of immersiveness, environmental fidelity, and sociability.*" The authors also list the key technological building blocks of the metaverse: networks, computing, 3D modeling, extended reality, the Internet of Things, blockchain, and artificial intelligence—which show the closeness and interdependence of 6G-based services and the metaverse discussions.

We envision the following 6G-enabled mobile operator business models:

- As an evolution of incumbent MNOs, the *6G MNO* business model will be building on end-to-end value chain controlled by the 6G MNO and supported by *specialized* firms tethered to the 6G MNO's connectivity-centered platform. This model will aim at monetizing interaction by 'matching' the needs or 'bridging' the customers via the connectivity platform. Automated network slicing will be used to offer differentiated service to segmented customer groups, private and public customers, and critical infrastructure providers. 6G MNOs are also expected to offer connectivity from a multi-technology platform that will consist of a selection of connectivity platforms that vary from low-earth-orbit, drone, and terrestrial 6G to hyper-local networks with a special focus on components and interfaces in the system. 6G MNOs will be designed to serve the masses, whether humans or machines. Additionally, 6G MNOs are envisioned to support the human and enterprise metaverses by providing the basic connectivity for them. For traditional MNOs, the network neutrality principle may constrain value capture in providing the long-tailed distribution of differentiated future services. 6G MNOs will rely on their existing infrastructure assets on top of which 6G will be built.

- *The OTT operator* business model will build on content that the over-the-top (OTT) service provider wants the end user to connect to. Connectivity, acquired from other types of operators, will be bundled as free or subsidized with content to provide a 'full service' that enables combined customer attraction and locked-in-based value creation with a focus on maximizing demand to monetize content. Generally, OTT refers to digital service providers that bypass the traditional MNO's network to deliver audio, video, and other media over the Internet, utilizing the possibly revisited net-neutrality principles, affordably expanding their reach to the bottom four billion. In this model, the OTT operator as a *platform owner* builds on its' own cloud platform and content, leveraging connectivity from other types of operators, preferably from bit-pipe operators, tailored for their needs, and will be able to benefit from its large customer base in 'bridging' between customers or 'sharing' contents mode. Any content that the *complementors*' offer can flexibly be added to platform-owning OTTs' offering in this model. Data and algorithms will play a central role in the functioning of this business model. OTT operators will focus on human users and are envisioned to be the consumer or enterprise metaverse providers.

- *The edge operator* business model will build on the openness of ecosystems and modularized technology to provide tailored localized or zone-specific connectivity, content, computing (i.e., use the available hardware to process data), and context services, and in multiple locations or zones to scale the service. Edge operators can be seen as the future versions of vertical service local or context service local operators of 5G who can specialize in serving customers at the edge of data and connectivity platforms. The key to understanding edge operators is their context-specificity. Depending on the type of customers and use cases they serve, they will develop specific sets of capabilities 'matching' customer needs and 'sharing' data or information between them while leveraging cloud infrastructure assets. Edge operators will drive the value chain in the edge application context and even create new revenue sources with hyper-local cloud infrastructure services with scalability, required availability, and almost unlimited flexibility. The diminishing value share of MNOs in edge cloud deals will trigger the number of private wireless deals of edge operators bypassing MNOs leveraging their infra-assets and creating a service layer to limit their value capture. Webscale born edge operators particularly are driving their successful transactional platform business model into new/adjacent domains where winning platforms cover innovation and transaction. The edge operators can be seen either as *complementors* or *disruptors*, depending on their role in the business ecosystems. The edge operators' business focus is on organizations and communities with either human or machine users. The edge operators may be expected to be the supporters of industrial metaverses.

- *The telco broker* business model offers connectivity or other resources or assets needed for mobile services. It can be seen to emerge from specialized data, artificial intelligence, and interface-control based services in the converging multi-platform ecosystem of future 6G. The telco brokers lean on the additionality of disruptors, complementors, and specialized service providers in the multi-platform ecosystem to match and bridge the differing needs and resources together by 'brokering.' The key to understanding the telco brokers' business model is the way they combine algorithms, ready-made components and existing resources, interfaces, and data residing in the multi-platform ecosystem for the needs of their customers. The telco brokers' business focus may take many forms and is primarily defined by the needs of organizations and communities. Telco brokers may serve any kind of metaverses as a service enabler.

The Pervasiveness of Platforms

The open questions of the 6G business models relate to what the assumptions and starting points are in 6G. Up to 5G, mobile network services have been defined and approached as top-down from networks toward the users, and most often from a technology-dominated point of view. However, with human-centricity that has been adopted as one of the starting points for future 6G, there might be a need and opportunity to define and approach 6G from the bottom up, from the user toward the network. This kind of conceptual and architectural disruption would require seeing the 6G network services as *focal to the user and user needs* rather than local or nation-wide—or representing a certain technology generation. This disruption would allow for a higher degree of business model innovation and variation in services within the ecosystem. Specifically, edge operators and telco brokers could benefit from this kind of approach as new value and spillover effects could be created in other than consumer-focused businesses.

Seen as platform-based, the 6G ecosystem may include new types of stakeholders, apart from the traditional MNOs and local operators, network constructors, system integrators, developer ecosystems, content owners and dealers, device, equipment, and technology vendors such as semiconductor technology vendors, operating system providers, application interface developers, or human–machine interface providers, cloud platforms and data centers and marketplaces prevalent already in 5G. These new types of stakeholders could include trust or security service providers, brokers of different resources like data, spectrum, or infrastructures, and digital twins, just to mention a few. It is also conceivable that the emerging human–machine interfaces may give rise to a new kind of service-centric and complementary service-flow business model:

- *The service-flow* business model integrates focally for the user, whether human or a machine—or a swarm of them—a set of on-demand services that the user needs ubiquitously regardless of location or connectivity provider. Future metaverses are examples of services requiring a service-flow business model in the background. The shift to cloud-based services has changed how enterprises purchase software and its development. Application developers have more control than before over what is being purchased. Companies build their products to make it easy for developers to adapt and shift their expensive top-down go-to-market motion to bottom-up product-led growth, where customers can easily try out the product and expand usage over time. A decentralized platform will distribute value between the players while open-source software will lower market entry barriers, promote interoperability, and expedite development cycles based on shared knowledge. These service-flow business models may require enhanced privacy and security via integrated trust-services, specified network capabilities and resources, specialized human-machine interfaces that replace traditional devices or equipment, and advanced AI capabilities. The service-flow business model will disrupt the other envisioned 6G-enabled mobile operator business models by shifting the focus from platform and infrastructure-centric offerings to human-centric service demand.

Discussion and Conclusions

The same way as with 5G, modifying the approach by Teece et al. (2022), we may find three types of business model and business ecosystem configurations in the 6G context: 1) the *vertical* supply-side incumbent connectivity platforms represented by the 6G MNOs, 2) the *horizontal* demand-side adjacent and content platforms represented by the OTT operators, and 3) the *oblique* multisided and multilayered newborn commerce platforms represented by the edge operators and telco brokers. These three groups are depicted in Fig. 6.2.

Miehé et al. (2022) analyzed strategies in relation to how complementors used connectivity to join existing ecosystems and by looking at whether they attract or replace stakeholders with a deepened or broadened value proposition. First, the vertical supply-side incumbent connectivity platforms, exemplified by the 6G MNOs, build on connectivity and cloud technologies with specialized partners that are tightly tethered to the connectivity platform with a deepened value proposal and

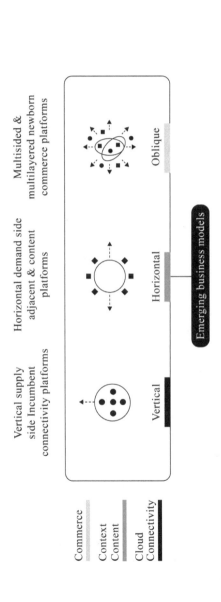

Fig. 6.2   Three types of business model and ecosystem configurations in 6G

with exploration strategy. Because it is connectivity-centric, this configuration aims to grow toward supporting content services or acting as a dealer of content such as media.

Second, the complementors of the horizontal demand-side adjacent and content platforms, exemplified by the OTT operators, broaden the value proposal of the configuration, with the aim to expand and exploit to new content areas, but also to new context businesses. The combination of content and context business models bundled with connectivity dealership creates a strong value proposition. In this configuration, the complementors are loosely coupled with the OTTs and may face fierce competition with other complementors.

Third, the oblique multisided and multilayered newborn commerce platforms, exemplified by both the edge operators and telco brokers, resemble the currently emerging division into service and tower companies in the telecommunications sector. Typical MNOs of today can be seen as focusing on services and may buy their network infrastructure from the 'tower' companies or infrastructure vendors that specialize in owning and managing the connectivity platform infrastructure and sell it as-a-service to the service operators. This configuration calls for a multitude of disruptions, but also complementary and specialized players, who may be in any area or combination of value creation with connectivity to cloud, content, context, or commerce business models. The service-flow business model is an example of the disruption of the whole mobile operator ecosystem, not only the new complementors' possible business models. Ecosystem stakeholders in this kind of configuration may have disruptive impacts on both up and downstream customer sectors as the traditional platform/infrastructure-centric ecosystem transitions via service-centricity to human-centric service flows. The vertical, horizontal, and oblique business models and their respective ecosystem structures coexist within the converging, multilayered multi-platform ecosystem of future 6G.

As a summary of the presented discussion, we may see the mobile communications business models as units of analysis to have developed *from technology-enabled and regulation-protected* national monopolies to *platform-based and regulation-delimited* international ecosystems. In parallel, we have witnessed the mobile communications market becoming more collaborative, ecosystemic, and having novel stakeholders serving transient positions and roles within the evolving ecosystem. In future, we can also expect the role of artificial intelligence to play a central role in assisting mobile communications and its use cases. As indicated in the discussion on platform business models, there is indeed a tendency in the extant research to see technology as a business model. However, business models need to be understood as devices used for sense making and commercializing technology—also future technology.

Bibliography

Ahokangas, P., Atkova, I., Yrjölä, S., & Matinmikko-Blue, M. (2023a). Business model theory and the becoming of mobile communications technologies. In A. Aagaard, & C. Nielsen, C. (Eds.), *BMI game changers*. Palgrave Macmillan.

Ahokangas, P., Matinmikko-Blue, M., Latva-aho, M., Seppänen, V. Arslan, A., & Koivumäki, T. (2021a). Future network operator business scenarios: Sharing economy and 5G. In R. Baikady, & P. Baerwald, (Eds.), *The Palgrave handbook of global social change*. Palgrave Macmillan.

Ahokangas, P., Matinmikko-Blue, M., & Yrjola, S. (2023b). Visioning for a future-proof global 6G from business, regulation and technology perspectives. *IEEE Communications Magazine, 61*(2), 72–78.

Ahokangas, P., Matinmikko-Blue, M., Yrjölä, S., & Hämmäinen, H. (2021b). Platform configurations for local and private 5G networks in complex industrial multi-stakeholder ecosystems. *Telecommunications Policy, 45*(5), 102128.

Ahokangas, P., Matinmikko, M., Yrjola, S., Okkonen, H., & Casey, T. (2013). "Simple rules" for mobile network operators' strategic choices in future cognitive spectrum sharing networks. *IEEE Wireless Communications, 20*(2), 20–26.

Al-Debei, M. M., & Avison, D. (2010). Developing a unified framework of the business model concept. *European Journal of Information Systems, 19*(3), 359–376.

Amit, R., & Zott, C. (2001). Value creation in e-business. *Strategic Management Journal, 22*(6–7), 493–520.

Barrera, K. G., & Shah, D. (2023). Marketing in the Metaverse: Conceptual understanding, framework, and research agenda. *Journal of Business Research, 155*, 113420.

Bennett, N., & Lemoine, G. J. (2014). What a difference a word makes: Understanding threats to performance in a VUCA world. *Business Horizons, 57*(3), 311–317.

Camps-Aragó, P., Delaere, S., & Ballon, P. (2019, September). 5G business models: Evolving mobile network operator roles in new ecosystems. *CTTE-FITCE: Smart Cities & Information and Communication Technology* (CTTE-FITCE), 1–6.

Chaharbaghi, K., & Lynch, R. (1999). Sustainable competitive advantage: Towards a dynamic resource-based strategy. *Management Decision, 37*(1), 45–50.

Chesbrough, H., & Rosenbloom, R. S. (2002). The role of the business model in capturing value from innovation: Evidence from xerox corporation's technology spin-off companies. *Industrial and Corporate Change, 11*(3), 529–555.

Chesbrough, H. (2010). Business model innovation: Opportunities and barriers. *Long Range Planning, 43*(2–3), 354–363.

Duan, Y., Fu, G., Zhou, N., Sun, X., Narendra, N. C., & Hu, B. (2015, June). Everything as a service (XaaS) on the cloud: Origins, current and future trends. *IEEE 8th International Conference on Cloud Computing*, 621–628.

Dahan, N., Doh, J., Oetzel, J., & Yaziji, M. (2010). Corporate NGO collaboration: Co-creating new business models for developing markets. *Long Range Planning, 43*(2–3), 326–342.

Dike, M. C., & Rose, E. L. (2017). Internationalization of mobile telecommunications: A systematic literature review. *Review of International Business and Strategy, 27*(3), 308–321.

Dike, M. C., & Rose, E. L. (2019). Cross-border expansion and competitive interactions of indigenous mobile network operators in sub-Saharan Africa. *Thunderbird International Business Review, 61*(1), 29–42.

Gawer, A., & Cusumano, M. A. (2014). Industry platforms and ecosystem innovation. *Journal of Product Innovation Management, 31*(3), 417–433.

Gooderham, P. N., Elter, F., Pedersen, T., & Sandvik, A. M. (2022). The digital challenge for multinational mobile network operators. More marginalization or rejuvenation? *Journal of International Management, 28*(4), 100946.

Hexa-X Project. (2021). *Deliverable D1.2. Expanded 6G vision, use cases and societal values.* https://hexa-x.eu/wp-content/uploads/2022/04/Hexa-X_D1.2_Edited.pdf.

Hmoud, A. Y., Salim, J., & Yaakub, M. R. (2020). Platformisation of mobile operators business model: A proposition using design science approach and grounded theory principles. *International Journal on Advanced Science Engineering Information Technology, 10*(2), 473–484.

Iivari, M. M., Ahokangas, P., Komi, M., Tihinen, M., & Valtanen, K. (2016). Toward ecosystemic business models in the context of industrial internet. *Journal of Business Models, 4*(2).

ITU-R. (2015). *IMT vision–framework and overall objectives of the future development of IMT for 2020 and beyond. Recommendation ITU, 2083*(0). https://www.itu.int/rec/R-REC-M.2083.

Ives, B., & Learmonth, G. P. (1984). The information system as a competitive weapon. *Communications of the ACM, 27*(12), 1193–1201.

Jahanbakht, M., Mostafa, R., & Veloso, F. (2022). Pre-entry experience, postentry adaptations, and internationalization in the African mobile telecommunications industry. *Organization Science, 33*(3), 969–990.

Kukliński, S., Tomaszewski, L., Kozłowski, K., & Pietrzyk, S. (2018). *Business models of network slicing.* 9th International Conference on the Network of the Future (NOF), 39–43.

Lehr, W., Queder, F., & Haucap, J. (2021). 5G: A new future for mobile network operators, or not? *Telecommunications Policy, 45*(3), 102086.

Li, F., & Whalley, J. (2002). Deconstruction of the telecommunications industry: From value chains to value networks. *Telecommunications Policy, 26*(9–10), 451–472.

Lindgren, P. (2016). Multi business model innovation in a world of 5G: What will persuasive business models look like in a world of 5G? *Wireless Personal Communications, 88*(1), 79–84.

Massa, L., Tucci, C. L., & Afuah, A. (2017). A critical assessment of business model research. *Academy of Management Annals, 11*(1), 73–104.

Matinmikko, M., Latva-Aho, M., Ahokangas, P., Yrjölä, S., & Koivumäki, T. (2017). Micro operators to boost local service delivery in 5G. *Wireless Personal Communications, 95*(1), 69–82.

Matinmikko, M., Latva-aho, M., Ahokangas, P., & Seppänen, V. (2018). On regulations for 5G: Micro licensing for locally operated networks. *Telecommunications Policy, 42*(8), 622–635.

Miehé, L., Palmié, M., & Oghazi, P. (2022). Connection successfully established: How complementors use connectivity technologies to join existing ecosystems–Four archetype strategies from the mobility sector. *Technovation,* 102660.

Morris, M., Schindehutte, M., & Allen, J. (2005). The entrepreneur's business model: Toward a unified perspective. *Journal of Business Research, 58*(6), 726–735.

Nambisan, S., Zahra, S., & Luo, Y. (2019). Global platforms and ecosystems: Implications for international business theories. *Journal of International Business Studies, 50,* 1464–1486.

Nielsen, C., & Lund, M. (2018). Building scalable business models. *MIT Sloan Management Review, 59*(2), 65–69.

Noll, J., & Chowdhury, M. M. (2011). 5G: Service continuity in heterogeneous environments. *Wireless Personal Communications, 57*(3), 413–429.

Rao, S. K., & Prasad, R. (2016). Telecom operators' business model innovation in a 5G world. *Journal of Multi Business Model Innovation and Technology, 4*(3), 149–178.

Rasheed, T., Radwan, A., Rodriguez, J., Kibilda, J., Piesiewicz, R., Verikoukis, C., … & Moreira, T. (2015). Business models for cooperation. In A. Radwan, & J. Rodriguez (Eds.), *Energy efficient smart phones for 5G networks* (pp. 241–267). Springer.

Rinaldi, F., Määttänen, H. L., Torsner, J., Pizzi, S., Andreev, S., Iera, A., … & Araniti, G. (2020). Non-terrestrial networks in 5G & beyond: A survey. *IEEE Access, 8,* 165178–165200.

Ritter, T., & Lettl, C. (2018). The wider implications of business-model research. *Long Range Planning, 51*(1), 1–8.

Sacoto-Cabrera, E. J., Guijarro, L., Vidal, J. R., & Pla, V. (2020). Economic feasibility of virtual operators in 5G via network slicing. *Future Generation Computer Systems, 109,* 172–187.

Snihur, Y., & Bocken, N. (2022). A call for action: The impact of business model innovation on business ecosystems, society and planet. *Long Range Planning, 55*(6), 102182.

Teece, D. J. (2010). Business models, business strategy and innovation. *Long Range Planning, 43*(2), 172–194.

Teece, D. J., Pundziene, A., Heaton, S., & Vadi, M. (2022). Managing multi-sided platforms: Platform origins and go-to-market strategy. *California Management Review, 64*(4), 5–19.

Timmers, P. (1998). Business models for electronic markets. *Electronic Markets, 8*(2), 3–8.

Wirtz, B. W., Pistoia, A., Ullrich, S., & Göttel, V. (2016). Business models: Origin, development and future research perspectives. *Long Range Planning, 49*(1), 36–54.

Wirtz, B. W., Schilke, O., & Ullrich, S. (2010). Strategic development of business models: Implications of the Web 2.0 for creating value on the internet. *Long Range Planning, 43*(2–3), 272–290.

Zhang, N., Cheng, N., Gamage, A. T., Zhang, K., Mark, J. W., & Shen, X. (2015). Cloud assisted HetNets toward 5G wireless networks. *IEEE Communications Magazine, 53*(6), 59–65.

Zott, C., & Amit, R. (2010). Business model design: An activity system perspective. *Long Range Planning, 43*(2), 216–226.

Zott, C., Amit, R., & Massa, L. (2011). The business model: Recent developments and future research. *Journal of Management, 37*(4), 1019–1042.

Yrjölä, S. S., Ahokangas, P., & Matinmikko-Blue, M. (2022). Value creation and capture from technology innovation in the 6G Era. *IEEE Access, 10*, 16299–16319.

Yrjölä, S. I., Ahokangas, P., & Matinmikko-Blue, M. (2021). Platform-based ecosystemic business models in future mobile operator business. *Journal of Business Models, 9*(4), 67–93.

Yrjola, S., Ahokangas, P., Matinmikko-Blue, M., Jurva, R., Kant, V., Karppinen, P., ... & Zepernick, H. J. (2020). *White paper on business of 6G* [White paper]. arXiv preprint arXiv:2005.06400.

7

Benefiting from Innovation in Future 6G

Pia Hurmelinna-Laukkanen and Seppo Yrjölä

To succeed, planning is insufficient. One must improvise as well.

(Isaac Asimov)

Benefiting from Innovation and General-Purpose Technologies

The profiting from innovation (PFI) framework (Teece, 1986) has widely been used to explain value capture from innovation. It explicates gaining full potential and capturing value from technological innovation (e.g., mobile communications industry) by innovators who utilize intellectual property protection provided by regulators (e.g. patents),

P. Hurmelinna-Laukkanen (✉)
Department of Marketing, Management and International Business, Oulu
Business School, University of Oulu, Oulu, Finland
e-mail: pia.hurmelinna@oulu.fi

© The Author(s) 2024   **167**
P. Ahokangas and A. Aagaard (eds.), *The Changing World of Mobile Communications*,
https://doi.org/10.1007/978-3-031-33191-6_7

inimitability afforded by the tacit nature of knowledge, and access to relevant complementary assets such as delivery channels.

However, considering the depth of the changes characterizing the contemporary innovation environment, the original PFI framework may have limited applicability and hence call for adjustment. As Gambardella et al. (2021, p. 75) suggest "*at least one important category of technology (i.e., enabling technologies) will not fit comfortably*" in the original PFI framework which is more coherent for explaining value capture from discrete inventions, such as the fourth-generation (4G) mobile communications technology providing connectivity to its users, with relatively narrow down-stream applicability (Kapoor & Teece, 2021). Enabling technologies are considered "*upgradable, adaptable technologies with improvement potential that have broad applicability which affects how returns accumulate from them*" (Gambardella et al., 2021, p. 75), like those emerging in the fifth generation of mobile communication technology (5G). Similar challenges are connected to general-purpose technologies (GPT), that is, technology characterized by general applicability (pervasiveness) across a variety of fields, technological dynamism, and innovation complementarities (Bresnahan & Trajtenberg, 1995). Examples of GPT have been noted to include steam power, electricity, the Internet, laser technology, and nanotechnology, as well as artificial intelligence (AI) (Yang et al., 2022), or 6G (Yrjölä et al. 2022).

To summarize the key differences between discrete, enabling, and general-purpose technologies and to link them with past, current, and

S. Yrjölä
Centre for Wireless Communications, University of Oulu, Oulu, Finland

Nokia, Oulu, Finland

S. Yrjölä
e-mail: seppo.yrjola@oulu.fi; seppo.yrjola@nokia.com

future mobile communications generations of 4G, 5G, and 6G, respectively, the following characterization is provided:

- Discrete technological solutions of 4G—innovation at the firm level—collaboration on the supply side—competition in the highly regulated demand (i.e., connectivity service) side.
- Enabling technologies of 5G—innovation at the ecosystem level—collaboration on both supply and demand sides—competition in the highly regulated demand side.
- General-purpose technology platforms/infrastructures of 6G—innovation at the ecosystem level between countries—collaboration and complementarities on both supply and demand sides—competition converging in multi-platform ecosystems comprising connectivity, data, and cloud platforms.

As the convergence of wireless and Internet technologies between industries proceeds and innovation increasingly takes place on platforms and in ecosystems driving the sixth generation of mobile communication networks (6G), the logic of gaining returns from innovative activities will also change. Importantly, the 6G context will be characterized by a completely new kind of convergence and new complementarities and there will be a central role for general-purpose technologies, which will generate new needs to understand how value can be captured from innovation. We propose here, following recent theorizing on innovation appropriability and appropriation (see, e.g., Gambardella et al., 2021), that instead of profiting from innovation, attention will turn to *benefiting from innovation*. We suggest that for 6G, a shift from the PFI approach to benefiting from innovation approach with interactive appropriability—i.e., "*context-specific dynamic aligning of appropriability premises (constituted with appropriability mechanisms and complementary assets) and interacting with other agents by relying on exclusion of others, leveraging the appropriability premises, and abandoning of protection, to benefit from innovation and appropriate value*" (Yang & Hurmelinna-Laukkanen, 2022, p. 311)—is relevant.

Building on the above discussion, we apply a general conceptual framework for innovation appropriability and appropriation. This framing provides a means for addressing the specificities of benefiting from innovation in the 6G context. Figure 7.1 provides an overview of the general logic of the approach.

The profiting from innovation framework looks especially into the logic of how innovators can monetize their innovation with the help of the so-called appropriability regime covering legal instruments (especially intellectual property rights, IPRs) and the (tacit, hard-to-imitate) nature of the technology, as well as *complementary assets* that give them bargaining power in the markets (Teece, 1986, 1998). Benefiting from innovation, however, requires a wider view. It considers not only a variety of appropriability mechanisms and complementary assets, but also how these instruments are utilized. It further addresses how such appropriability premises and their uses are aligned with(in) the contextual and situational factors surrounding them (Hurmelinna-Laukkanen & Yang, 2022; Yang et al., 2022). Likewise, this view explicitly acknowledges that the appropriation outcomes may not be only about the immediate profits from innovation (see Ahuja et al., 2013, about primary appropriability), but the benefits may be quite varied from private to social returns, and they may accrue over time. These elements are central components of the *interactive appropriability* (Yang & Hurmelinna-Laukkanen, 2022) that embraces the dynamism and investigates the alignment of the appropriability-enhancing *instruments*, the *processes* in which they are used, influential contextual factors, and appropriation *outcomes*, as depicted in Fig. 7.1.

Applying the Benefiting from Innovation Framework to 6G

6G—which can be considered an emerging general-purpose platform—will transform how and what kinds of services are offered, responding simultaneously to increasing societal demands for resilience, sustainability, inclusion, and empowerment (Yrjölä et al., 2022). This means that the dimensions of appropriability and appropriation will have their own specific nature in this context.

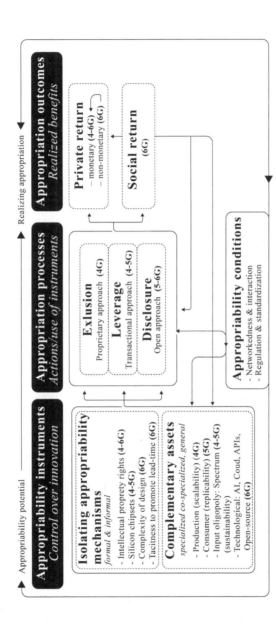

Fig. 7.1 Benefiting from innovation in a 6G context framework with a comparison to 4G and 5G (*Source* Hurmelinna-Laukkanen and Yang, [2022])

Appropriability Mechanisms and Complementary Assets

First, in the 6G context, isolating appropriability mechanisms and complementary assets for GPT may well be the same as those for more discrete technologies or in other innovation contexts. Intellectual property rights (IPRs), contracts, tacitness, lead time, and human resources-related *isolating appropriability mechanisms* may be utilized, and brand and sales assets may be acquired as relevant complementary assets to support value capturing (see Hurmelinna-Laukkanen & Puumalainen, 2007). However, these may be affected by the appropriability conditions, which makes the selection of appropriability mechanisms and complementary assets in 6G context more complex and may influence the emphasis given to individual instruments.

In fact, certain shifts can be observed regarding the selection of appropriability mechanisms. In 4G (and 5G), relying on IPRs was the norm, and complementary assets such as access to distribution channels or marketing capabilities were highlighted. Complexity and high external transaction costs drove mobile network operators to own and operate network assets, especially spectrum, radios, tower, backhaul, data centers, computers, storage, and transport. At present, the Internet and information technologies, coupled with financial and supply chain sophistication, enable mobile operators to exploit assets owned by others such as towers, cloud data centers, computing and storage resources, which has already turned attention to the role of consumers and replicability as well as input oligopoly, for example, as complementary assets that emerge as opportunities (and varying tensions) in the networked settings where the selection of the appropriability instruments calls for careful attention (see Hurmelinna-Laukkanen & Nätti, 2012; Stefan et al., 2021).

The future decentralized 6G, especially with Web3 blockchain and smart contracts, is expected to dramatically lower the cost of external (market) transactions which will encourage further disaggregation to maximize business agility, improve resource utilization, and better align business risks and rewards. Standards in this area and various regulatory requirements for maintaining market competition, handling data, and ensuring security, for example, will influence how firms can and need to approach appropriability instruments. Business and regulatory restrictions may constrain value appropriation from downstream sectors (Yrjölä et al., 2022). In addition to telecommunications regulations and related competition laws, challenges arise from information technology regulation and industry segment-specific regulations. For example, the network neutrality principle may constrain value capture in providing the long-tailed distribution of differentiated future services (Frias & Martínez, 2018). Critical issues in 6G will be regulation related to multi-sided digital cloud platforms and the governance of the privacy and security of users, especially affecting the data protection and artificial intelligence rights. Furthermore, technological innovations may require access to new technological public or semi-public *complementary assets* such as the authorization of the radio spectrum, access regulation of the obligations for interoperability, and the use of public infrastructure (Yrjölä et al., 2022).

Spectrum authorizations, including administrative allocation, market-based mechanisms, and the unlicensed commons approach, will play a key role in defining the wireless markets and ultimately defining who can operate various wireless systems (Matinmikko-Blue et al., 2020). Timely access to affordable quality spectrum resources will be fundamental in allowing new market entry for innovative wireless solutions and a powerful tool for incentivizing and forcing different spectrum users to act toward sustainability goals desired by the national regulators (EC RSPG21-027, 2021). In fact, a relevant complementary asset could be seen to stem from the need to have suitable knowledge assets to handle these demands; access to (or even ownership of) regulatory and standardization related knowledge is increasingly important.

Exclusion, Leverage, and Disclosure

Second, the 6G context will be inherently characterized by open innovation logic, which will directly influence the appropriation processes. Whereas 4G can be considered to have been focused on exclusion and proprietary strategies of the central players, the business models discussed in the 5G literature highlight more connectivity service provisioning and its differentiation (Yrjölä et al., 2022), which reflects an increasing reliance on leverage processes (Fig. 7.1).

6G, on the other hand, has been envisioned as an intelligent system of systems that will converge connectivity with complementary services such as imaging, sensing, and location, opening numerous new application areas and business opportunities (Latva-aho & Leppänen, 2019). The emerging 6G business ecosystems will be characterized with novel resource configurations and changing and novel stakeholder roles such as local operators, cloud operators, infrastructure providers, and resource brokers. The openness of business models boils down to discussions on open innovation (Chesbrough, 2003, 2006), and in platform contexts, this brings the ecosystem and its stakeholders close. For example, a software-based, service-oriented cloud-native 6G network could enable efficient infrastructure and resource sharing by different tenants, and could open the ecosystem to new players, and accelerate the time to market by reducing service creation and activation times.

Modularization and open interfaces facilitate competition and entry, enabling stakeholders to specialize within the ecosystem and develop complements to the platform. Consequently, start-ups can increasingly access complementary assets through various forms of alliance with larger firms. Thus, 6G as a GPT may have large positive innovation spillovers and externalities, both stable and dynamic, that alter the valuation of existing technologies and enable opportunities for third-party and novel stakeholders (Carlaw & Lipsey, 2002). Against this setting, the appropriation processes are highlighted; actors need to be highly selective in terms of exclusion, leverage, and disclosure-oriented approaches (see Hurmelinna-Laukkanen & Yang, 2022). Employing the full range of alternative appropriability instruments (isolating appropriability mechanisms and complementary assets) is likely to be needed in the 6G context, where situations may change quickly. Especially for GPT, openness may be imperative in the search for best use cases (Yang et al., 2022), but this does not imply that no benefit could be gained by keeping some elements away from the reach of others. For example, while modular architecture with open interface specifications will enable the majority of software functions to be deployed on any commercial computing hardware, lower-layer processing-intensive radio functions may continue to be implemented in custom silicon chipsets with proprietary technological innovations.

Private and Social Returns

Finally, it needs to be acknowledged that the wanted (and pursued) outcomes matter already at the stage of strategizing on and building of appropriability premises. While the original PFI framework considers private return to the innovator to be an integral incentivizing factor (Teece, 1986, 1998), the changes brought by technological developments and shifts toward platform and ecosystem-based approaches have augmented the view to include appropriation outcomes (Hurmelinna-Laukkanen & Yang, 2022). In the 6G context, these aspects have increased their importance. Appropriability may include the appropriation of social value (at different levels from self-reward for individuals to wide benefits that match the United Nations Sustainable Development Goals (SDGs)), which could be highly relevant in contemporary innovation environments where sustainability and other such aspects have increased in importance. 6G is no exception to this, and mobile communication networks, increasingly underpinning mission-critical functions across communities and businesses after 2030, will not only transform how and what kind of services are offered; they will also be shaped by the growing societal requirements of resilience, inclusion, sustainability, and empowerment (Yrjölä et al., 2022).

Social value is not necessarily where the benefit from innovation ends. In Fig. 7.1, the arrows from social return to private return illustrate this. Alnuaimi and George (2016) have noted that knowledge retrieval can benefit original innovators later even if they release (or lose) the inventive technologies to the surrounding environment where others exploit it. Spillovers and social returns may become a source of private returns for the initial innovator especially in case of GPT (see Yang et al., 2022). In the 6G context, this could happen via increased use of as-a-service (aaS) business models where resources and infrastructure are not purchased, but consumed as a service (Matinmikko et al., 2021). For example, serverless cloud application development and execution models will allow developers to build and run code and applications without managing servers, and without paying for idle cloud infrastructure.

Conclusions and Implications

The above discussion suggests that interactive appropriability (Yang & Hurmelinna-Laukkanen, 2022)—resonating with value co-capture within the 6G ecosystem—will be highly relevant in order to benefit from innovation in the 6G context. In particular, it advocates the relevant players to combine firm- and network-points of view in a meaningful way. Innovators can and need to carefully align their selection of appropriability instruments with the prevailing conditions—paying specific attention to the regulatory and other such requirements, and the networked context in which they operate (see Hurmelinna-Laukkanen & Nätti, 2012). At the same time, they need to align the use of those instruments with the pursued appropriation outcomes. Importantly, the relatively more notable role of social returns compared to private returns requires acceptance of uncertainty, instead of immediate private return. At the societal level, the largest rewards for 6G as a GPT may come from later, yet unforeseeable activity. In the 6G context, some examples, and signals of this are already visible, exemplified by the opportunity to strengthen prosumerism and human-centricity via digital inclusion and rural mobile communications. As a part of its enabling role, 6G will help different sectors of society to monitor and renew sustainable operations via solutions and services combining communication with other services such as imaging, sensing, and locating, with hyper-local granularity.

It appears evident that 6G will co-exist with several generations of mobile communications technologies, which indicates of the importance of backward compatibility, continuity toward 7G, and complementarities between technology domains over the evolution of the generations. To date, this evolutionary development has followed the define-standardize-develop/deploy-use cycle of technology commercialization (Ahokangas et al., 2023). The benefit from an innovation framework embracing the logic of interactive appropriability, therefore, has direct implications for innovation policies applied to developing 6G as a GPT, as well as implications for regulatory authorities framing future 6G, and firms and entrepreneurs aiming at commercializing 6G.

Implications for Innovation Policy

Currently, global competition to define future 6G has begun. Digital technologies in general have become the "*battleground for the competition for global leadership*" (Moerel & Timmers, 2021, p. 5). This can be seen in the launches of national 6G research and innovation programs, and the collaboration between countries, for example, at the European level. As innovation policies have direct and indirect impacts on a firm's innovation practices and intellectual property creation, in the case of GPTs, international cross-industry innovation sets new demands for integrated, transformative, and directed innovation policies based on a shared vision of what 6G could become. To allow innovators to benefit from 6G innovation, the specific attention of the innovation policies should, on the one hand, be targeted to create a favorable environment for appropriability instruments, i.e., isolating mechanisms and complementary assets, to work, especially in AI. On the other hand, innovation policies should pay attention to social returns and spillover effects on upstream and downstream industries and the society at large. Transformative innovation policies (TIPs) may help developing converging and complementary new technologies (Bailey et al., 2019). A TIP in the context of 6G as a GPT could mean an innovation policy based on vision-based directed innovation, a twin focus on covering market imperfections and failures, and new emerging opportunities for complementarities and creating a competitive edge and new markets.

Implications for Regulatory Authorities

In today's mobile communications context, the national regulatory authorities define rules of the business by allowing, limiting, or directing what kind of activities are possible for the ecosystem stakeholders. Traditionally, the regulators' key concern has been to ensure competition and innovation in the markets, depending on the local market conditions and available technologies. However, given the increasingly complex regulatory domain affecting the development, deployment, and use of 5G and future 6G, regulatory authorities would be advised to consider the appropriability conditions emerging for 6G to maximize the national potential for appropriability in both 5G and 6G. A good question is what the compound impact of regulations on the appropriability of 6G will be. Although the appropriation outcomes, private and social rents, have been of interest for regulators, it is also expected that the role of societal rents and implications are becoming central in the 6G context, especially due to the increasing role of AI for 6G. It already is a fact that global 5G and 6G will be facing different regulatory logics combined with varying innovation policies and market approaches. The market-based US, the rights-based European, and China's government push-based logic in developing new technologies will have an impact on how these technologies generate private and societal rents for different firms and societies (Feijóo et al., 2020).

Implications for Firms and Entrepreneurs

Finally, the benefiting from innovation framework can be highly valuable for firms attempting to benefit from 6G innovation. For incumbent firms, the changing role of innovation and ecosystems will mean the need to develop new ecosystem-embracing strategies and business models in their existing markets. The incumbents should pay attention to emerging new ecosystems and business verticals in the intersection of different technological domains, considering the relevance of interactive appropriability. Similarly, start-ups and small and medium-sized enterprises should explore growth opportunities with complementary services in emerging 5G and 6G ecosystems from this perspective; while smaller firms tend to focus on immediate financial return for obvious reasons, they also need to be aware of the search for varying appropriation outcomes in the ecosystems around them. The discussed framework shows how appropriability instruments may be available to all firms, but that the readiness to engage in varying appropriation processes may open new insights leading to benefits from innovation in the emerging GPTs—also other than 6G.

Bibliography

Ahokangas, P., Atkova, I., Yrjölä, S., & Matinmikko-Blue, M. (2023). Business model theory and the becoming of mobile communications technologies. In A. Aagaard & C. Nielsen (Eds.), *BMI game changers*. Palgrave MacMillan.

Ahuja, G., Lampert, C. M., & Novelli, E. (2013). The second face of appropriability: Generative appropriability and its determinants. *Academy of Management Review, 38*(2), 248–269.

Alnuaimi, T., & George, G. (2016). Appropriability and the retrieval of knowledge after spillovers. *Strategic Management Journal, 37*(7), 1263–1279.

Bailey, D., Glasmeier, A., Tomlinson, P. R., & Tyler, P. (2019). Industrial policy: New technologies and transformative innovation policies? *Cambridge Journal of Regions, Economy and Society, 12*(2), 169–177.

Bresnahan, T. F., & Trajtenberg, M. (1995). General purpose technologies 'Engines of growth'? *Journal of Econometrics, 65*(1), 83–108.

Carlaw, K. I., & Lipsey, R. G. (2002). Externalities, technological complementarities and sustained economic growth. *Research Policy, 31*(8–9), 1305–1315.

Chesbrough, H. (2003). *Open innovation—The new imperative for creating and profiting technology*. Harvard Business School Press.

Chesbrough, H. (2006). *Open business models: How to thrive in the new innovation landscape*. Harvard Business School Press.

EC RSPG21-027. (2021). *Draft RSPG opinion on the role of radio spectrum policy to help combat climate change*. EC Directorate-General for Communications Networks, Content and Technology, Radio Spectrum Policy Group.

Feijóo, C., Kwon, Y., Bauer, J. M., Bohlin, E., Howell, B., Jain, R., Potgieter, P., Vu, K., Whalley, J., & Xia, J. (2020). Harnessing artificial intelligence (AI) to increase wellbeing for all: The case for a new technology diplomacy. *Telecommunications Policy, 44*(6), 101988.

Frias, Z., & Martínez, J. P. (2018). 5G networks: Will technology and policy collide? *Telecommunications Policy, 42*(8), 612–621.

Gambardella, A., Heaton, S., Novelli, E., & Teece, D. J. (2021). Profiting from enabling technologies? *Strategy Science, 6*(1), 75–90.

Hurmelinna-Laukkanen, P., & Nätti, S. (2012). Network orchestration for knowledge mobility: The case of an international innovation community. *Journal of Business Market Management, 5*(4), 244–264.

Hurmelinna-Laukkanen, P., & Puumalainen, K. (2007). Nature and dynamics of appropriability: Strategies for appropriating returns on innovation. *R&D Management, 37*(2), 95–112.

Hurmelinna-Laukkanen, P., & Yang, J. (2022). Distinguishing between appropriability and appropriation: A systematic review and a renewed conceptual framing. *Research Policy, 51*(1), 104417.

Kapoor, R., & Teece, D. J. (2021). Three faces of technology's value creation: Emerging, enabling, embedding. *Strategy Science, 6*(1), 1–4.

Latva-aho, M., & Leppänen, K. (Eds.) (2019). *Key drivers and research challenges for 6G ubiquitous wireless intelligence*. http://jultika.oulu.fi/files/isbn97 89526223544.pdf.

Matinmikko-Blue, M., Yrjölä, S., & Ahokangas, P. (2020). Spectrum management in the 6G era: The role of regulation and spectrum sharing. In *2020 2nd 6G wireless summit* (6G SUMMIT) (pp. 1–5). IEEE.

Matinmikko-Blue, M., Yrjölä, S., Ahokangas, P., Ojutkangas, K., & Rossi, E. (2021). 6G and the UN SDGs: Where is the connection? *Wireless Personal Communications, 121*(2), 1339–1360.

Moerel, L., & Timmers, P. (2021). *Reflections on digital sovereignty*. https://euc yberdirect.eu/research/reflections-on-digital-sovereignty.

Stefan, I., Hurmelinna-Laukkanen, P., & Vanhaverbeke, W. (2021). Trajectories towards balancing value creation and capture: Resolution paths and tension loops in open innovation projects. *International Journal of Project Management, 39*(2), 139–153.

Teece, D. J. (1986). Profiting from technological innovation: Implications for integration, collaboration, licensing and public policy. *Research Policy, 15*(6), 285–305.

Teece, D. J. (1998). Capturing value from knowledge assets: The new economy, markets for know-how, and intangible assets. *California management review, 40*(3), 55–79.

Yang, J., Chesbrough, H., & Hurmelinna-Laukkanen, P. (2022). How to appropriate value from general-purpose technology by applying open innovation. *California Management Review, 64*(3), 24–48.

Yang, J., & Hurmelinna-Laukkanen, P. (2022). Benefiting from Innovation— Playing the Appropriability Cards. In: R. Agarwal, R. Green, E. Patterson, &

S. Pugalia (Eds.), *Innovation* (Chapter 15, pp. 310–331). Routledge. https://doi.org/10.4324/9780429346033-20.

Yrjölä, S., Ahokangas, P., & Matinmikko-Blue, M. (2022). Value Creation and Capture from Technology Innovation in the 6G Era. *IEEE Access, 10,* 16299–16319.

Part III

Regulatory and National Considerations

8

Local 5G/6G Network Business in Europe: Regulatory Analysis and Legitimacy Considerations

Oxana Gisca, Marja Matinmikko-Blue, Petri Ahokangas, Seppo Yrjölä, and Jillian Gordon

The intersection of law, politics, and technology is going to force a lot of bright thinking.

(Bill Gates)

O. Gisca (✉) · P. Ahokangas
Martti Ahtisaari Institute, Oulu Business School, University of Oulu, Oulu, Finland
e-mail: oxana.gisca@oulu.fi

P. Ahokangas
e-mail: petri.ahokangas@oulu.fi

M. Matinmikko-Blue
Infotech Oulu Focus Institute and Centre for Wireless Communications, University of Oulu, Oulu, Finland
e-mail: marja.matinmikko@oulu.fi

S. Yrjölä
Centre for Wireless Communications, University of Oulu, Oulu, Finland

© The Author(s) 2024
P. Ahokangas and A. Aagaard (eds.), *The Changing World of Mobile Communications*,
https://doi.org/10.1007/978-3-031-33191-6_8

The Challenge of Local Networks

Local 5G networks have gained attention in recent years, and different stakeholders have started to establish local and often private mobile communication networks within specific settings, such as factories (Ahokangas et al., 2021; Matinmikko et al., 2018). This kind of local 5G networks may be expandable to a variety of location and context-specific use cases in locations such as hospitals, campuses, shopping malls, and mass event arenas or to accommodate the distinct business-related requirements of various sectors such as the automotive, media, entertainment, health, utility, and industry vertical sectors and the rated opportunities for tailored offerings (Ahokangaset al., 2022). Indeed, several key business opportunities for local 5G operators have been identified including offering hosted local connectivity to different mobile network operators (MNOs) in specific locations resulting in a neutral host model, and/or providing secure local networks for vertical-specific needs with locally tailored services resulting in private networks (Matinmikko et al., 2017). The local 5G/6G networks offer connectivity to local data, improved service quality, and privacy and security assurances (Ahokangas et al., 2019). Security, privacy, reliability, and the management of local data of 5G/6G networks are essential for businesses.

As the next step, the development of 6G as a new general-purpose technological platform is increasingly being framed by new tensions to innovate the entire business ecosystem (Yrjölä et al., 2021, 2022). These tensions can be explained as resulting from different policy areas, including the need to protect users' privacy, security, and safety, whether

Nokia, Oulu, Finland

S. Yrjölä
e-mail: seppo.yrjola@oulu.fi; seppo.yrjola@nokia.com

J. Gordon
Adam Smith Business School, University of Glasgow, Glasgow, Scotland, UK
e-mail: jillian.gordon@glasgow.ac.uk

they be people or machines (Yrjölä et al., 2020), as well as from sharpened innovation and competition policies (Van Duijvenvoorde, 2020) and fragmented regulatory developments (Ahokangas, 2022).

Regulating Local Mobile Communication Networks

Different countries have employed different strategies related to local 5G/6G networks (Cave, 2018). The difficulties in developing a reliable local 5G/6G network business require a multidisciplinary approach, involving economic, regulation, and technology concerns. Since local 5G/6G networks respond to some already identified business security concerns this emphasizes a range of emerging challenges alongside the security and privacy ones. Local 5G/6G secure mobile connectivity for multimedia communications and content represents an option for ensuring better security. The cybersecurity of 5G and 6G must, therefore, be improved (Bauer, 2022) so that people and businesses can trust, use, and benefit from innovation in connectivity and increased automation. Additionally, it is important to protect fundamental human rights and freedom including the right to privacy and the protection of personal data, as well as the freedom of expression and information.

However, that would not be sufficient since the emergence and commercial success of local private networks in 5G/6G depend on their recognition and acceptance. In short, the emergence and success depend on legitimacy (Suchman, 1995) in the eyes of the different stakeholders of the mobile communications industry/ecosystem (Marano et al., 2020; Thomas & Ritala, 2022), and specifically the regulatory authorities (Ahokangas et al, 2021; Yrjölä et al., 2022). Legitimacy can be seen as a 'proxy-indicator' for assessing the complex institutional dynamics that influence the embedding of a new industry in relevant structures (Bergek et al., 2008). Moreover, a solid regulatory framework is imperative for the establishment of local 5G/6G networks to ensure competition, innovation, and the emergence of new services. Thus, the process that is needed to gain legitimacy for local private 5G/6G networks is not only a multifaceted regulatory challenge but also a business challenge impacting the business models and ecosystems needed to provide connectivity services in local settings securely. In this, a specific challenge is the legitimation of local private network services that are offered by non-MNOs in the mobile communications ecosystem (Matinmikko-Blue et al., 2021).

The Legitimation of Local Mobile Communication Networks

Regulations and regulators act as enablers for the legitimacy that currently the 5G/6G local network are lacking. The literature mentions that there is an essential lacking component associated with the regulatory frame of future localized networks. As a disruptive and developing innovation, private local 5G networks face several 'industry legitimacy' or 'industry acceptance'-related issues (Kwak & Yoon, 2020; Marano et al., 2020), which must be taken into account. Regulatory approval should aim to not only mitigate the risks, but also strike a proper balance between defining risk and benefits, developing effective mechanisms for proper regulation of local 5G/6G networks business and promoting innovation. In Europe, the current landscape for telecommunication regulation is evolving at the EU level (Bauer & Bohlin, 2022).

Telecommunication regulation in the EU member states is quite complex and undertaken on the national, European, and international levels. The EU regulatory framework for electronic communications is formulated as a set of policy objectives, which national regulatory agencies implement with the help of instruments defined at the European level. The member states are currently implementing the provisions of the European Electronic Communications Code (EECC) (EU, 2018a) established in 2018, relying on related hard law and soft law legal provisions. Regarding the legitimacy of local private mobile communication networks, one of the challenges is the lack of definition for this concept in the EU legal act. More importantly, newly developed initiatives such as the Digital Services Act (EU, 2022b) and the Digital Market Act (EU, 2020a) are silent about their specific features influencing local private networks. At the same time, new regulations such as those governing the use of artificial intelligence are being introduced, potentially influencing local private networks and related services. What regulatory solutions can be developed for local 5G/6G networks operated by different stakeholders are currently on the agenda of the countries that aim to promote innovation in the context of mobile communication.

Aims of the Chapter

Analyzing the landscape of the EU legal framework of (secure) electronic communications, the chapter proposes a new, legitimation-based approach for understanding and analyzing relevant regulations for local private 5G/6G networks, especially focusing on non-MNO entities. Decisions to start providing local private mobile communications services are fundamentally business decisions, made under the prevailing regulatory conditions and based on the available enabling technologies in the business ecosystem. The success of these new businesses is influenced by the legitimacy received from the institutional environment (EU, 2018b). Creating a new business with the new technology can thus be seen as a way of testing the legitimacy of that technology, provided that the regulation allows or supports it.

Building on the above argumentation, this chapter presents an assessment of the most relevant EU legal developments and underlines the legitimacy challenges for local private 5G/6G networks. It aims to frame the EU telecommunication legal framework relevant for local 5G networks and presents future perspectives to be developed by policymakers. While identifying that the upcoming 5G/6G local networks will have a large societal impact and that the regulation will enable their future deployment, there is very little prior research on regulatory challenges or the ecosystem legitimacy of local 5G/6G networks. Therefore, this chapter aims to address the following research questions:

- What comprises the legal framework and its elements which are relevant to local 5G/6G networks in Europe?
- How could the emerging legitimacy challenges of local 5G/6G networks be addressed?

- Which managerial choices and consequences impact the legitimacy of new local 5G/6G network businesses?

Our analysis follows a thematic analysis, which is described as a method for identifying, analyzing, and reporting pattern themes within data (Braun & Clarke, 2006). In general, the analytical interventions used in a thematic analysis cover the identification of the themes, reviewing the themes, and searching for themes that are the units of analysis (Elo & Kyngäs, 2008). A theme is defined as a coherent integration of the disparate pieces of data that constitute a finding. By applying the thematic analysis method, we want to contribute by comprehending, ordering, and revealing the implications to extant research (Alvesson & Sandberg, 2020). As an emerging phenomenon the local private networks require a novel approach to reframe the problem-field, re-order the elements that are relevant to it, and to support sense-making.

The chapter proceeds as follows. We first discuss the context and the key concepts of the research, then we continue with thematic analysis of the key regulations related to local networks, and close with the regulatory and managerial implications of our analysis.

The Emerging 5G/6G Local Mobile Communication Network Business

Digital technologies are transforming our world. Information and communication technologies (ICTs) have an essential role in this transformation and have a major impact on practically every aspect of society. The traditional mobile communication business value chain has been changing incrementally. However, mobile network operators' (MNOs) market dominance has been shaken by the Internet giants that offer over-the-top (OTT) services that have reduced MNOs' role to operating as bit pipes (Matinmikko-Blue et al., 2019).

The development of 5G networks has expanded from the traditional MNO-centric deployment models to alternative local network operator models. Local 5G operators are emerging on the mobile market, offering local high-quality services in high-demand spatially confined locations, such as factories, sports arenas, and campuses (Matinmikko et al., 2017). Moreover, the 5G technology is expected to further change the mobile communications market structures, by addressing different vertical sectors' specific local service demands. This market development challenges the traditional MNO dominance and progressively opens the market to new business opportunities and new stakeholders, including the local private network service providers.

There is growing interest in the vertical sectors toward the deployment of their own local 5G networks tailored for specific service delivery without being tied to the existing MNOs. The use of 5G within vertical sectors, such as manufacturing, has particularly attracted recent attention and regulators have already taken actions to promote the spectrum assignment decisions in several countries, such as Finland and Germany, and many others are considering it. How the spectrum allocation and assignment decisions will influence the emerging local deployment and new operational models is linked to nations' competitiveness through the growing digitalization in all aspects of society. This also shapes the business opportunities for many companies. Previous research on 5G local mobile communication network (Matinmikko et al., 2019) showed that the MNO market dominance has continued in many countries with the early 5G spectrum decisions, and only a subset of countries have allowed market entry for local and often private 5G networks by introducing local spectrum licensing (see Matinmikko et al., 2018). For regulators, there is a choice of how to balance promoting efficient spectrum use, fairness, competition, and innovation in spectrum allocation. Currently, different countries have different goals for their spectrum policies and employ different strategies for local networks. At the same time, the spectrum awarding decisions taken by the regulators in the new 5G bands have varied between different countries significantly (see e.g. Matinmikko et al., 2019; European 5G Observatory, 2021). As the spectrum and competition regulation have played a pivotal role in allowing, delimiting, or protecting/safeguarding certain business models applied by the operators, technology-oriented business studies have been complemented with research on regulation and policy as an antecedent for new business opportunities (Yrjölä et al., 2022).

From the local operator's perspective, a diverse set of service offerings can emerge, ranging from serving MNOs' customers as a neutral host to operating private networks for specific verticals with different revenue potential and models. Recent studies on regulatory developments and legitimacy for local 5G networks (Matinmikko et al., 2018) have identified several key regulatory elements to be considered by policymakers including access regulation, consumer protection, competition enforcement and economic regulation, intellectual property, privacy and data protection, resource management, network security, taxation, and universal service and accessibility. The future development of 5G technology and markets is undetermined and regulations must be developed with incomplete knowledge and under conditions of uncertainty. Such conditions call for an adaptive policy (Bauer & Bohlin, 2022). Therefore, monitoring the experience at the national and international level of the regulatory developments will facilitate global learning and help improve the efficiency of the policy framework. The ecosystem identity, which means a *"set of mutual understandings among ecosystem participants regarding central, enduring, and distinctive characteristics of the ecosystem value proposition"* (Thomas & Ritala, 2022, p. 3), is essential to ecosystem legitimacy attainment and value realization.

Key Concepts of Research

As an emerging innovation, private local 5G/6G networks may share several 'industry legitimacy' challenges (Kwak & Yoon, 2020; Marano et al., 2020). Regarding the existing legal provisions in force, some regulatory-related challenges that need to be considered have been identified in recent studies (Matinmikko et al., 2018). In general terms, legitimacy can be seen to mean that the converged connectivity and data platform constellations are considered appropriate for and accepted by the industry's relevant stakeholders (Kwak & Yoon, 2020; Suchman, 1995). To successfully legitimize a new venture such as a private local network in an emerging new industry, managers will have to change and/ or create a new structural meaning of norms, practices, and values for it (Turcan & Fraser, 2016). Legitimacy in the industrial context means the consonance of an industry with its institutional environment (Kwak & Yoon, 2020) and can be defined as the ability to select the 'right thing to do' (Palazzo & Scherer, 2006).

Regulatory interventions influence emerging businesses directly and indirectly, impacting their legitimacy. The regulatory design provides important boundary conditions to route legitimacy. The definition of legitimacy as a right thing to do (Palazzo & Scherer, 2006) can be interpreted from the mobile communications business perspective as action-oriented choices focused on the available opportunity, value creation and capture, and opportunities for (competitive) advantage. Similarly, the consequences of these actions should be that the business is scalable, replicable, and sustainable (Yrjölä et al., 2022). An emergent ecosystem can be considered legitimate by ecosystem participants and other actors in the broader environments when the key managerial choices regarding business opportunities, value creation and capture, and competitive advantages and consequences of the mentioned choices, the scalability, replicability, and sustainability, are covered or achieved. With the definition by Adner (2017), who described ecosystems as "*the alignment structure of the multilateral set of partners that need to interact in order for a focal value proposition to materialize*," it becomes clear that value propositions represent an integral and central part of an ecosystem (Adner, 2017, p. 40).

According to the recent studies on the ecosystem (Phillips & Ritala, 2019), when such organizational forms are emerging, they require legitimacy to overcome the 'liability of newness.' Adopting a collective action lens and taking legitimacy as a process approach, a process model of ecosystem collective action has been proposed, where orchestrators, complementors, users, and external actors together rule ecosystem legitimacy. Within this research, we consider legitimacy as a process focusing on the aspects that lead to the emergence of legitimacy (Suddaby et al., 2017). The symbiosis of actions based on the business thinking approach in regulatory and business actions may lead to ecosystem legitimacy.

The business model has become the contemporary paradigm for innovating revealing about business and researching firm behavior in increasingly dynamic business environments. The business model is perceived as a tool for conducting boundary-spanning analysis in contemporary business research (Lanzolla et al., 2020; Zott et al., 2011). Business models are made of concrete choices and the consequences of these choices. Scholars contend that they are composed of two different sets of elements: the concrete choices made by management about how the organization must operate and the consequences of these choices, in addition to the effects on value creation and/or value capture different designs have different specific logics of operation and create different value for their stakeholders (Casadesus-Masanell & Ricart, 2010). Scholars distinguish three types of choices: policies, assets, and governance structures. Policy choices refer to courses of action that the firm adopts for all aspects of its. Asset choices refer to decisions about tangible resources. Governance choices refer to the structure of contractual arrangements that confer decision rights over policies or assets. The three types of choices may be depicted from the business constructs as well as for regulatory mechanisms.

The business model can be perceived as a driver for creating competitive advantage through opportunity exploration and exploitation. Given the contemporary business environment, an advantage is rarely sustainable and can be quickly rendered uncompetitive (McGrath, 2010). A business model analysis also gives us a sense of firms in action. Advances in ICT and the demands of societally motivated enterprises constitute important sources of recent business model innovations (Casadesus-Masanell & Ricart, 2010). As the theoretical focus of this research, the symbiosis between managerial choices, opportunity, value, advantages, and consequences—scalability, replicability, and sustainability—in the business ecosystem and regulatory actions for the purpose of legitimacy attainment will be analyzed.

Thematic Analysis of Relevant Regulations

Adopting a business-orientated approach toward the regulatory measures reflected in the EU documents developed under the Digital Single Market Strategy (EU, 2015), we depicted nine EU legal acts from the overall EU regulatory system. Particularly, these include the following: (a) the EU 2030 Digital Compass (EU, 2021b); (b) the Digital Markets

Act (EU, 2020a); (c) the Digital Services Act (EU, 2022b); (d) the Data Act (EU, 2022a); (e) the Data Governance Act (EU, 2020b); (f) The Cybersecurity Strategy (EU, 2022c); (g) the E-privacy Regulation (EU, 2017); (h) the Directive (EU), 2018/1972 of the European Parliament and of the Council establishing the European Electronic Communications Code (EU, 2018a); and (i) the AI Act (EU, 2021a) as depicted in Fig. 8.1.

Analyzing the specific relevant legal provisions that may be developed further in the national context and tailored toward the deployment of local 5G/6G networks, we depicted the managerial choices and consequences for emerging 5G/6G businesses. In the thematic analysis of the EU Digital Legal Framework, we established the key elements and then

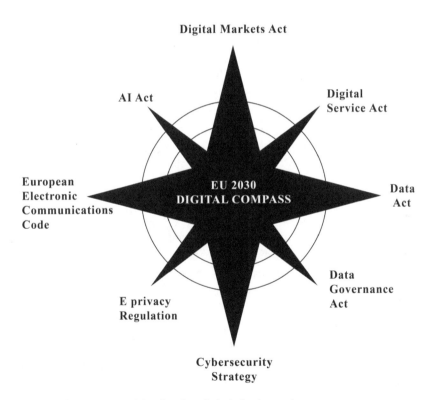

Fig. 8.1 The EU's priorities for the digital single market

prioritized the business needs for 5G/6G networks to develop a broad understanding of future regulatory settings. Moreover, relying on the regulatory and business perspectives, the analysis proposes for discussion of challenges and future perspectives that policymakers may undertake at the national and international level that pose significant weight for legitimacy attainment. New insights into the studied phenomenon informed practical action (Krippendorff, 2004). The process comprises three steps. First the identification of the relevant regulatory landscape and its key elements for local and private networks for 5G/6G. Second the assessment of the managerial implications in terms of managerial choices and consequences, and third, the discussion of the consequent regulatory challenges and perspectives.

The European Union's Governments the European Parliament and European Commission have agreed on key regulations that will overhaul the EU digital market. The European Commission aspires to make the EU's single market fit for the digital age, moving from the 28 national digital markets to a single market. The digital single market opens new opportunities, as it removes key differences between online and offline worlds, breaking down the barriers to cross-border online activity and moving beyond technology frontiers. The Digital Single Market Strategy (EU, 2015) was adopted in May 2015, and it is made up of three policy pillars:

- *An environment where digital networks and services can prosper.* The digital single market (EU, 2015) aims to create the right environment for digital networks and services by providing high-speed, secure, and trustworthy infrastructures and services supported by the right regulatory conditions. Key concerns of the single market include cybersecurity, data protection/e-privacy, and the fairness and transparency of online platforms,
- *The digital technology as a driver* for growth pillar aims at maximizing the growth potential of the European digital economy so that every European can fully enjoy its benefits, notably by enhancing digital skills, which are essential for an inclusive digital society, and

● *Improving access to digital goods and services.*

The strategy seeks to ensure better access for consumers and businesses to online goods and services across Europe, e.g., by removing barriers to cross-border e-commerce and access to online content while increasing consumer protection. The review of the Telecom Framework is one of the pillars of the EU Digital Single Market Strategy. As one of the EU political priorities, the Digital Single Market Strategy (EU, 2015) echoes the aspirations toward synchronizing European values from the physical to the digital world.

Figure 8.1 presents the identified legal instruments and provides a comprehensive structure of the main legislative initiatives that were identified as connected (Seretschy, 2021) via the key objectives and emphasizes another viewpoint from the three-dimensional path outlined by this research. EU legislators are focused on creating more effective regulation enforcement, creating a more flexible regulatory environment, and even new, future regulatory frameworks. Allowing businesses to bring a product to market more quickly under existing regulatory conditions, or by testing out adaptations to existing regulations, an iterative, flexible regulatory system may be developed. This will also help obtain a nuanced understanding of a technology's impact on businesses. However, it is difficult to predict the effects of a technological change until it has become widely adopted. However, once a technology has become entrenched in business and social practices, changing its effects would be difficult. Ex-ante regulation legislation aims to resolve the conflict between the lumbering legislative process and the rapidly evolving technology.

Identified Legal Instruments

This section presents the identified regulatory developments with an impact and increased relevance for local 5G/6G networks, discussing the objectives, aim, and key content of the regulations and highlights the stakeholders' obligations.

EU 2030 Digital Compass

In March 2021, the European Commission presented a vision and avenues for Europe's digital transformation by 2030—the Digital Compass for the EU's digital decade (EU, 2021b). The Digital Compass aims to empower citizens and businesses with a human-centric, sustainable vision for a digital society. The Digital Compass evolves around four cardinal points or aims: *Government* for the digitalization of public services, *Skills* in terms of developing digitally skilled population and highly skilled digital professionals, creating secure and performant sustainable digital *Infrastructures*, and ensuring digital transformation of *Businesses*. The Commission will first create the anticipated EU trajectories for each goal before assessing how well the EU member states are doing in achieving these goals. Each Member State would also specify national anticipated trajectories and offer national strategic roadmaps that outline their coordinated actions to meet the objectives. Every year, the progress will be evaluated along with the national and EU trajectories. The Digital Compass provides the following obligations:

- Putting people and their rights at the center of the digital transformation
- Supporting solidarity and inclusion
- Ensuring the freedom of choice online
- Fostering participation in the digital public space
- Increasing safety, security, and empowerment of individuals
- Promoting the sustainability of the digital future

Digital Markets Act

In March 2022, a political agreement was reached on the Digital Markets Act (DMA) (EU, 2020a) that aims to make the digital market in Europe more transparent, safe, and accountable. It aims to promote fair competition in digital markets and give SMEs (small and medium-sized firms) a chance to participate better in the data economy by fostering innovation, growth, and competitiveness, and facilitate the scaling up of smaller platforms, small and medium-sized enterprises and start-ups under a clear framework at EU level and preventing gatekeepers from imposing unfair conditions on businesses and end users and at ensuring the openness of important digital services.

The DMA concerns the largest online platforms, social networks, search engines, online marketplaces, advertising services, among others. The new regulations, which address a range of digital challenges in the less digitally developed economies, might have an impact on the entire planet. The DMA would grant the European Commission new enforcement powers that could influence the business models of major Internet corporations because it will oversee enforcing the compliance of 14 digital gatekeeper platforms with 21 new competition laws. With the DMA, Europe is setting standards for how the digital economy of the future will function. The European parliamentarian Andreas Schwab (EPP, DE) mentioned in an EU press release in 2022 that the law avoids any form of overregulation for small businesses. "*App developers will get completely new opportunities, small businesses will get more access to business-relevant data and the online advertising market will become fairer.*" (European People's Party, 2022). The DMA establishes clearly defined obligations vis-à-vis a very limited number of cross-border providers of core platform services:

- Transparency
- Due diligence
- Prohibition of unfair practices
- Data portability & interoperability
- Access for business users

Digital Services Act

To complement the DMA, agreement on a common set of rules was reached in April 2022 on intermediaries, for example online marketplaces, social networks, content-sharing platforms, app stores, and online travel and accommodation platforms, and their obligations and accountability across the single market that aim to open new opportunities to provide digital services across borders while ensuring a high level of protection to all users. The proposed Digital Services Act (DSA) (EU, 2022b) aims to set common but tailored obligations and accountability rules for providers of network infrastructure, cloud computing services (such as Internet access providers), hosting service providers, and particularly for online platforms (i.e., online marketplaces, and social media platforms). For the first time, full-fledged oversight and enforcement rules are envisaged with the ability to set fines of up to 6% of the global annual turnover of platforms. According to the law, *"digital services are a broad category of online services, from straightforward websites to services for internet infrastructure and online platforms"* (EU, 2022b). Regardless of where the company is located, all digital services that operate in the EU are subject to the Digital Service Act, including small and micro-businesses; albeit the requirements are adapted to firm size. 90% of the impacted enterprises in the EU are small to medium-sized businesses that will be spared from the most expensive requirements. The DSA is fundamentally a legislative framework that will set guidelines for how platforms must manage their content, marketing, and how they apply algorithmic techniques. It strengthens the responsibilities and supervision of intermediary service providers to ensure less citizen exposure to illegal content and products online, contributes to the proper functioning of the internal market for intermediary services, and set out uniform rules for a safe, predictable, and trusted online environment, where fundamental rights enshrined in the act are effectively protected.

The key stakeholders of the DSA include intermediary service providers, social networks, online marketplaces, and hosting services. The DSA sets obligations for:

- Transparency
- Due diligence
- Content moderation
- Risk management
- Online advertising rules

Data Act

The proposed regulation on ten harmonized rules on fair access to and use of data, the Data Act (DA) (EU, 2022a), makes an important contribution to the digital transformation objective of the Digital Decade. It is a key measure for making more data available for use in line with EU rules and values for business-to-business and in some cases business-to-government transactions. The DA creates a mechanism to enable the safe reuse of given categories of public-sector data that are subject to the rights of others. The DA establishes guidelines for the use of data produced by Internet of Things (IoT) devices, ensuring fairness in the allocation of data value among actors within the data economy. It covers the technical, economic, and legal problems that result in underutilization of data. The DA specifies who can use data to generate value and under what circumstances. The DA addresses a wide range of organizations, from service providers and gatekeepers to device makers and governmental agencies, and the key ramifications and prospects for European firms are related to aims for data exchange and access requirements. Cloud switching, interoperability standards, and data sharing are all impacted by the Data Act (EU, 2022a). The DA is highly relevant for SMEs as data portability requirements allow shifting between services.

The key stakeholders of the DA comprise product manufacturers and service suppliers in the EU, data holders and data recipients in EU, public bodies and EU institutions, and providers of data processing services, setting the obligations for the:

- Access and use of non-personal data
- Data portability, facilitating switching, and interoperability
- The fair, reasonable, and non-discriminatory approach in data sharing contracts
- Public usage of data

Data Governance Act

In May 2022, the European Council approved a new law to promote the availability of data and build a trustworthy environment to facilitate their use for research and the creation of innovative new services and products. The Data Governance Act (EU, 2020b) represents a robust mechanism to facilitate the reuse of certain categories of protected public-sector data, increase trust in data intermediation services with third countries, and foster data altruism for common good across the EU. The act is an important component of the European strategy for data, which aims to bolster the data economy. The Data Governance Act (EU, 2020b) complements the 2019 Open Data Directive (EU, 2019a), which does not cover such types of data, and creates a framework to foster a new business model—data mediation services—that will provide a secure environment in which companies or individuals can share data.

For businesses, these services can come in the form of digital platforms that encourage voluntary data sharing between businesses and make it easier to comply with the data-sharing requirements imposed not only by this law but also by other legislation, whether it be at the national or European level. By utilizing these services, businesses may share data without worrying about it is being abused or losing its competitive edge. Providers of data intermediation services must be listed in a register. The Data Governance Act (EU, 2020b) also makes it simpler for people and businesses to voluntarily make data available for initiatives for public good like research projects and innovation.

The key stakeholders of the DGA include the public sector, data sharing "trust" services, and citizens. The DGA sets obligations for confidentiality and one-stop shop mechanism for data requests.

EU Cybersecurity Strategy and Cybersecurity Act

The EU's cybersecurity strategy (EU, 2020c) and the related Cybersecurity Act (EU, 2019b) aim to boost Europe's collective resilience against cyber threats and help to ensure the fundamental rights that all citizens and businesses can fully benefit from trustworthy and reliable services and digital tools including open global Internet. These exist to fortify efforts for secure digitalization and promote norms for world-class solutions and standards of cybersecurity for essential services and critical infrastructures and drive the development and application of new technologies. The key stakeholders for the Cybersecurity Strategy include key services and infrastructures like energy, transport, banking, financial market infrastructures, health, drinking water, wastewater, and digital infrastructures as well as public administration and space EU.

Under the new Cybersecurity Strategy (EU, 2020c), Member States are urged to finish implementing the EU 5G Toolbox, a thorough and unbiased risk-based strategy for the security of 5G and future generations of networks, with the help of the Commission and ENISA, the European Cybersecurity Agency. The Commission's Recovery Plan for Europe, the Security Union Strategy 2020–2025, and Shaping Europe's Digital Future are all important parts of the new EU Cybersecurity Strategy (EU, 2020c) for the Digital Decade. The obligations of the Cybersecurity strategy include:

- Addressing both cyber and physical resilience of critical entities and networks
- Reforming the rules on the security of network and information systems
- Boosting cybersecurity and matching EU level investment
- Completing the implementation of the EU 5G Toolbox, a comprehensive and objective risk-based approach for the security of 5G and future generations of networks

E-Privacy Regulation

The regulations for e-privacy is intended to provide up-to-date rules and procedures, for instance, for the increasingly popular messenger services—and to do so uniformly throughout the EU. The e-Privacy Regulation (ePR) (EU, 2017) is a proposal for the regulation of various privacy-related topics, mostly in relation to electronic communications within the European Union. It lays down rules regarding the protection of fundamental rights and freedoms of natural and legal persons in the provision and use of electronic communications services. It ensures free movement of electronic communications data and electronic communications services within the EU member states. The ePR specifies what forms of electronic information enjoy its protection and how businesses can use such data. It introduces rules on cookies, direct marketing, and business-to-business communications and will replace the outdated e-Privacy Directive from 2002. It also aims to increase user security, the confidentiality of communication and metadata, to define clearer rules for technologies such as cookies, and control of spam.

The key stakeholders of the ePR include electronic communications services, electronic communications software providers, and natural and legal persons who use user-related information to send marketing messages. It sets obligations for:

- Adjusting data anonymization
- Rules for the use of metadata
- Access to information on end-user devices

The European Electronic Communications Code

The directive (EU, 2018a) puts in place one of the essential building blocks for a digital single market in Europe (EU, 2015). The EU Directive marks a significant revision of the regulatory framework, dating from 2009. It creates a legal framework to ensure freedom to provide electronic communications networks and services. It represents the main legal act which aims to stimulate competition and increase investment in 5G and very high-capacity networks (fixed and mobile) so that every citizen and business in the EU can enjoy high-quality connectivity, a high level of consumer protection, and an increased choice of innovative digital services. It also aims to develop an internal market across the EEA and ensure the protection of consumers. The directive introduces a series of new objectives and tasks: strengthened consumer rules aim to make it easier to switch between service providers while offering better protection. Basically, the EEEC responds to the increasing convergence of telecommunications, media, and information technology so that all electronic communications networks and services should be covered to the extent possible by a single European electronic communications code.

The key stakeholders of the EEEC include various electronic communications networks, electronic communications services, over-the-top internet players, and public bodies and EU institutions. The EEEC sets obligations for:

- Providing operators with predictable regulation
- Ensuring there is no discrimination between network and service providers operating under similar circumstances
- Applying the regulation in a technology-neutral fashion whenever possible and relevant rules facilitating new market entrants
- Implementing the "use it or lose it principle" with respect to the withdrawal of spectrum licenses
- Facilitating a shared use of mobile frequencies

AI Act

The proposal for harmonized rules in the Artificial Intelligence Act (AIA) (EU, 2021a) amend certain EU legislative acts. With the AIA, the Commission aims to address the risks generated by specific uses of AI through a set of complementary, proportionate, and flexible rules. It applies to all sectors (except for the military), and to all types of artificial intelligence. These rules will also provide Europe with a leading role in setting the global gold standard. The legal framework for AI proposes an approach based on three risk categories. First, applications and systems that create an unacceptable risk, such as government-run social scoring of the type used in China, are banned. Second, high-risk applications, such as a CV-scanning tool that ranks job applicants, are subject to specific legal requirements. Lastly, applications not explicitly banned or listed as high-risk are largely left unregulated. Like the European Union's General

Data Protection Regulation (EU, 2016), the AI Act (EU, 2021a) could become a global standard. It is already having impact beyond Europe by:

- Ensuring that AI systems are safe and respect fundamental values
- Creating legal certainty to facilitate investment and innovation in AI
- Addressing the risks stemming from the various uses of AI systems and promotes innovation in the field of AI

The key stakeholders of the AIA include providers placing AI systems on the EU internal market, users of AI systems within the EU, and the providers and users of AI in a third country when outputs are used in the EU. The AIA sets obligations for:

- Ex-ante risk assessments
- Respect for fundamental rights
- Transparency toward users
- Post-market monitoring and reporting
- Human oversight

Discussion and Conclusions

This chapter set out to examine three research questions related to the local 5G/6G networks. The first question—What comprises the legal framework and its elements relevant to local 5G/6G networks in Europe?—was answered by the review of telecommunication provisions presented in the previous section, covering the main pillars of the EU Digital Single Market Strategy influencing local 5G/6G networks. The second question—How could the emerging legitimacy challenges of local 5G/6G networks be addressed?—we see as related to the regulatory framework in general, and it will be discussed in detail in the subsequent section. The third question—What managerial choices and consequences impact legitimacy of new local 5G/6G network businesses? —we see to concern the local 5G/6G businesses directly and will be examined in the last part of this section.

Regulatory Challenges and Future Policy Considerations

The cross-cutting impact of digital innovation and data has already removed old sector boundaries and created new legitimacy challenges for emerging sectors. Emerging businesses have removed old system boundaries, directly challenging old practices. In response to these changes, we observe the need for the emergence of proactive, future-oriented, and innovation-enabling approaches to regulation in Europe. We are beginning to see a change in the theory of regulation with the emergence of a new field as 'ex-ante' regulation or 'anticipatory' but practice lags behind. The new approach to regulation helps reframe regulation as a supportive tool for the responsible development and use of new technologies and business models. New and existing methods are helping regulators to better support innovation as it emerges, drive innovation directly, and respond faster or act preemptively to prevent public harm.

This study identified regulatory challenges and particular elements relevant for 5G/6G local networks from the EU Digital Legal Framework that pose legitimacy challenges that may be considered further. Figure 8.2 shows the identified regulatory challenges and perspectives for legitimacy attainment.

The EU legal framework provisions ensuring security within and between the networks and coordination and control across multiple locations do not cover in detail the specific features and main characteristics of 5G/6G local mobile communication networks, especially for those not deployed by the MNOs. However, security and privacy are the main elements that need to be considered in all the network installations for local 5G/6G network deployment. Clear and tailored provisions for vertical-specific industries concerning managing and ensuring the confidentiality of data ownership or exploitation of personal data are not defined and developed in the EU legal framework, which may have a cascading effect on the quality of implementing measures by the NRAs at the national level. Current spectrum management awarding mechanisms are designed for wide area MNO networks and do not properly address the emerging local 5G/6G networks.

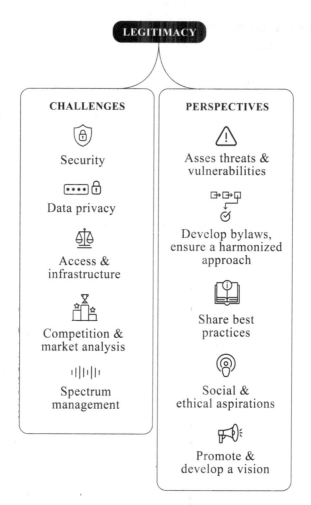

Fig. 8.2 Regulatory legitimacy challenges and perspectives

One of the core elements vital to the emergence of 5G/6G local networks are market analyses and competition provisions. Clear and transparent delineation of the relevant markets and firms present in multiple-related markets were not reflected in the analyzed content. The lack of a market monitoring framework and developed criteria pose obstacles for emerging of 5G/6G local network. Our analysis emphasizes future perspectives for regulators and business in ensuring a qualitative step beyond, for legitimacy attainment. First and foremost, promotion and developing a vision on local 5G/6G networks to disseminate knowledge and promote a positioning for acceptance among companies and other relevant stakeholders needs a prioritization approach. Building best practices and common instruments and facilitating the harmonization of implementation of legal provisions would be an efficient tool for tackling regulatory and business challenges. To elaborate and share best practices and tailored implementation mechanisms to encourage the active participation of the NRA is a perspective that needs to be treated with proper allocation of resources at the EU level. Ensuring a harmonized approach of the NRAs in developing bylaws and guidelines for implementation of EU directives with the support of EU monitoring bodies needs to be expanded. Additionally, promoting a social and ethical approach via the government and public–private–people (PPP) programs are on the agenda of some governments that promote innovation. Sharing best practices in this will provide efficient insights for national regulators on the complex regulatory and business landscape.

This chapter has identified and provided an analysis of the complex regulatory landscape that impacts the emerging local 5G/6G networks, depicting its relevant elements. It characterizes the connection to business opportunities around the current local 5G networks and upcoming local 6G networks and identifies regulatory challenges and perspectives for legitimacy attainment. The analysis results indicate that reviewing the evolution of the European digital framework from a legitimacy attainment perspective following the incorporated European values is a powerful way of illustrating the shift of the parameters of the regulatory in promoting European standards. Analyzing the EU legal framework relevant to electronic communications, it proposes a new approach for considering the legitimacy of emerging 5G/6G local network regulation. Also, the business thinking approach helps technology adoption, promotes regulation supporting experimentations, and considers systemic effects of policy and regulation.

The chapter also analyzed the potential effects of identified regulatory interventions that are currently in place or under consideration. In contrast to earlier research, the discussion relies on an approach inspired by ecosystem legitimacy. As mentioned by Boyd (2000), legitimacy is the result of a process and can be defined as a strategy effectuated by an organization. New business models claim legitimacy to grow, expand, and exist. Ecosystem legitimacy could represent a way for business models to be more active in establishing and promoting a competitive system that encourages the adoption of emergent business models throughout the ecosystem and ensure efficient private and public cooperation.

Following the previous research, we may assume that regulatory interventions influence the emerging business models directly and indirectly. The regulatory design provides important boundary conditions to guide legitimacy and represents an enabler for innovation. Opportunities, value creation and capture, and advantages as managerial choices and scalability, replicability, and sustainability as managerial consequences provide an overarching approach for guiding the policymakers in developing and streamlining the regulatory actions.

Applying the novel theoretical approach of ecosystem legitimacy developed in this study, the in-depth analysis identifies the EU legal acts that are relevant for local 5G/6G networks business consideration. Moreover, the analysis provides systematized key concepts, actors, and obligations that are under consideration of the policymakers agenda. It identifies that security, data privacy, spectrum management access and infrastructure, and competition and market analysis are regulatory elements that may pose challenges for legitimacy attainment. Future perspectives for stakeholders were presented and discussed. Ensuring, developing, and promoting a vision will increase social political support and will raise awareness of 5G/6G local networks. Building and sharing best practices will boost the transparency of regulatory mechanisms and will assist NRA to promote clear and efficient implementation mechanisms. Developing and ensuring a harmonized approach via bylaws and sectorial legislation will ensure that the core provisions are properly implemented. Ensuring social and ethical aspirations via human-centric regulations will support awareness on sustainability and trustworthiness. Adopting a proper mechanism for assessing vulnerabilities will determine the allocation of resources in network access and infrastructure.

Although the proposed theoretical approach has reflected on the legitimacy challenges arising from the EU Legal Digital Framework overall, certain regulatory actions and provisions may require extra, context-specific variables when determining such challenges. As legitimacy is an audience-dependent construct, certain stakeholders and audiences might have specific needs that might have been overlooked within the selected proposed framework and will be addressed in some further work. Additional research into how to facilitate the process of legitimation of 5G/6G local network is needed particularly from the stakeholder's perspective. In the same vein, discussions on national sovereignty not only for critical infrastructures but the all ICT technologies have become a concern, raising the need for new kinds of governance and regulation. In the current political and economic climate, which is evidently more open to industrial policy considerations, and in which, regarding local 5G/6G network business, some countries have routinely engaged in some form of industrial intervention, despite their varied economic choices. Further discussion with a focus on a regulatory approach should not be delayed any longer. Regulators should consider handling both ex-ante and ex-post mechanisms: ex-ante when local 5G/6G networks are designed and ex-post when they are implemented. Future research needs to be conducted on the development and application of the proper regulatory mechanisms for the widespread adoption of local 5G/6G networks businesses.

Managerial Implications Related to the Legitimacy of Local 5G/6G Networks

With the emergence of new technologies and business models, policymakers face the question of whether the existing legal and regulatory framework is appropriate, or whether a different market design might more fully realize the potential benefits for society. Our analysis indicates that the regulatory landscape of mobile communications networks, especially concerning the case of local private networks in 5G/6G, has changed and includes now new areas to be considered. Clearly, a new and holistic approach is needed to make sense of the changes in the regulatory landscape. Following the thematic analysis of the EU identified legislative pillars for digital ecosystem provisions, the EU Digital

Legal Framework, we identify a set of *managerial choices* related to new opportunities, value creation/capture, and advantages that regulations should consider for local 5G/6G networks. Similarly, we identify a set of *managerial consequences* related to scalability, replicability, and sustainability of local 5G/6G networks. Table 8.1 presents some examples of managerial choices and consequences for MNOs and non-MNOs regarding 5G/6G local network businesses that arise from the analyzed EU Digital Legal Framework.

The EU Digital Legal Framework *does not* contain definitions or specific rights or obligations for local 5G/6G networks for offering services. However, the EU Digital Legal Framework may trigger several managerial choices as new opportunities are expected to emerge based on new digital communication features such as high-precision holographic media or digital senses provided over the network. For example, gigabit network connectivity will allow value creation by meeting various vertical-specific industry demands. SMEs in manufacturing could have local access to cloud-based innovative industrial service platforms such as manufacturing-as-a-service systems and market places to boost the efficiency of their production capacities. Within healthcare, health data and records could be processed quickly locally, and in agriculture the deployment of edge capabilities connected to farming machinery would allow collecting data in real time and could provide advanced services to farmers for harvest prediction or farm management and the optimization of food supply chains. Especially, the areas labeled as strategic areas under the provisions provided by the 2030 Digital Compass may present various business opportunities for companies. New choices may be related to remote application servers that will be easier to access, bring more information content, utility applications, and realistic forms of communication directly to the consumer via edge clouds.

The uptake of digital solutions and the use of data will trigger managerial consequences related to transitioning to a climate-neutral, circular, and more resilient economy. As highlighted in the European

Table 8.1 Managerial choices and consequences derived from the identified legal framework for emerging local 5G/6G private networks

Choices Opportunities, Value, Advantages	Legal frame-Work	Consequences Scalability, Replicability, Sustainability
Relevant for 5G local mobile communication networks deployed by all	**Digital Compass** **Digital Market Act** **Digital Service Act** **Data Act** **Data Governance Act** **Cyber-security Strategy** **e-Privacy Act** **EECC** **AI Act**	*Relevant for 5G local mobile communication networks deployed by all*
• Develop use cases and expand in priority vertical ecosystems		• Align business objectives with social, economic, and ecological sustainability, trust, the digital green transition, fair and responsible utilization of data, digital inclusion and equality
• Incorporate edge computing		• Develop a strategy to clarify the legal conditions for AI adoption
• Monetize 5G technology for higher data rates, lower latency, and massive device density		*Relevant for 5G local mobile communication networks deployed by MNOs*
• Participate and cooperate in data-sharing		• More alternative technology vendors
• Benefit from better access to data		• Increased competition
• Facilitate data flows through technical standards and interoperability		• International digital partnerships under the digital decade strategic framework are expected to become more common
• Develop new data-based business opportunities		• Coopetition strategies (parallel competition and collaboration) may become an efficient tool for developing business
• Benefit from transparent procedures of data portability		*Relevant for 5G local mobile communication networks independent of MNOs*
• Take advantage of data sharing in common goods and services		• A stand-alone, isolated network under the spectrum management system may become an efficient option
• Create a secure and trusted online environment		• Sharing of data between the partners may become more efficiently secured and faster
• Apply to risk-aware mechanisms in adopting and promoting AI		• Capital and operational expenses may be minimized by using simple radios, the edge cloud, and the core network at the edge of the cloud

(continued)

Table 8.1 (continued)

Choices Opportunities, Value, Advantages	Legal frame-Work	Consequences Scalability, Replicability, Sustainability
Relevant for 5G local mobile communication networks deployed by MNOs • Expand cloud infrastructure • Benefit from standardization efforts		
Relevant for 5G local mobile communication networks independent of MNOs • Activate a more secure business ecosystem • Apply for rights to use spectrum • Benefit from network slicing, interoperability and interconnection		

Data Strategy,[1] the volume of data generated due to increasing digitalization is greatly increasing and a growing proportion of data is expected to be processed at the edge, closer to the users and where the data is generated. This shift will require the development and deployment of fundamentally new data processing technologies encompassing the edge and moving away from centralized cloud-based infrastructure models. These new trends will result in new managerial consequences related to increasing distribution and decentralization of data processing capacities and suitable supply of cloud services. Seeking replicability of solutions, managers are expected to draw on partnership development under the legal provisions. Acting in a clear risk aware environment and stable regulatory conditions will enable resource allocation and planning for reaching scalable business decisions. We may conclude that the adoption of a business-oriented approach in regulation—that considers both managerial choices and consequences—would enable easier deployment of local and private networks in 5G/6G and facilitate their legitimation throughout the 5G/6G ecosystem.

Acknowledgements The authors would like to acknowledge LNETN-project from the European Union's Horizon 2020 research and innovation program under the Marie Skłodowska-Curie grant agreement No 860364 and 6G Flagship program at the University of Oulu (Grant No. 318927).

Bibliography

Adner, R. (2017). Ecosystem as structure: An actionable construct for strategy. *Journal of Management, 43*(1), 39–58.

Ahokangas, P. (2022). An action plan for benefiting from European innovation in future mobile connectivity. In *Europe's future connected: Policies and challenges for 5G and 6G networks*. European Liberal Forum.

Ahokangas, P., Matinmikko-Blue, M., Latva-aho, M., Seppänen, V., Arslan, A., & Koivumäki, T. (2022). Future mobile network operator business

---

[1] https://commission.europa.eu/strategy-and-policy/priorities-2019-2024/europe-fit-digital-age/european-data-strategy_en.

scenarios: Sharing economy and 5G. *The Palgrave handbook of global social change* (pp. 1–25). Springer International Publishing.

Ahokangas, P., Matinmikko-Blue, M., Yrjölä, S., & Hämmäinen, H. (2021). Platform configurations for local and private 5G networks in complex industrial multi-stakeholder ecosystems. *Telecommunications Policy, 45*(5), 102–128.

Ahokangas, P., Matinmikko-Blue, M., Yrjölä, S., Seppänen, V., Hämmäinen, H., Jurva, R., & Latva-aho, M. (2019). Business models for local 5G micro operators. *IEEE Transactions on Cognitive Communications and Networking, 5*(3), 730–740.

Alvarez, F., Breitgand, D., Griffin, D., Andriani, P., Rizou, S., Zioulis, N., Moscatelli, F., Serrano, J., Keltsch, M., Trakadas, T. P., Phan, K., Weit, A., Acar, U., Prieto, O., Iadanza, F., Carrozzo, G., Koumaras, H., Zarpalas, D., & Jimenez, D. (2019). An edge-to-cloud virtualized multimedia service platform for 5G networks. *IEEE Transactions on Broadcasting, 65*(2), 369–380.

Alvesson, M., & Sandberg, J. (2020). The problematizing review: A counterpoint to Elsbach and Van Knippenberg's argument for integrative reviews. *Journal of Management Studies, 57*(6), 1290–1304.

Anker, P. (2017). From spectrum management to spectrum governance. *Telecommunications Policy, 41*(5–6), 486–497.

Ballon, P. (2009). The platformisation of the European mobile industry. *Communications & Strategies, 75*, 15.

Bauer, J. M. (2022). A framework for 5G and 6G market design. *Erik Bohlin Francesco Cappelletti*, 20.

Bauer, J. M., & Bohlin, E. (2022). Regulation and innovation in 5G markets. *Telecommunications Policy, 46*(4), 102260.

Bauer, J. M., & Bohlin, E. (2019). *The role of regulation in 5G market design* (Quello center working paper, TPRC47). 47th Research Conference on Communication, Information and Internet Policy.

Bergek, A., Jacobsson, S., & Sandén, B. A. (2008). 'Legitimation' and 'development of positive externalities': Two key processes in the formation phase of technological innovation systems. *Technology Analysis & Strategic Management, 20*(5), 575–592.

Boyd, J. (2000). Actional legitimation: No crisis necessary. *Journal of Public Relations Research, 12*(4), 341–353.

Braun, V., & Clarke, V. (2006). Using thematic analysis in psychology. *Qualitative Research in Psychology, 3*(2), 77–101.

Casadesus-Masanell, R., & Ricart, J. E. (2010). From strategy to business models and into tactics. *Long Range Planning, 43*(2–3), 195–215.

Cave, M. (2018). How disruptive is 5G? *Telecommunications Policy, 42*(8), 653–658.

Corporate Europe Observatory. (2022). *Big tech's last-minute attempt to tame EU tech rules.* https://corporateeurope.org/en/2022/04/big-techs-last-min ute-attempt-tame-eu-tech-rules.

Elo, S., & Kyngäs, H. (2008). The qualitative content analysis process. *Journal of Advanced Nursing, 62*(1), 107–115.

EU. (2015). *Communication on a digital single market strategy for European Commission to the European Parliament, the Council, the European Economic and Social Committee and the Committee of the Regions.* https://eur-lex.eur opa.eu/legal-content/EN/TXT/?uri=celex%3A52015DC0192.

EU. (2016). *General data protection regulation.* https://gdpr-info.eu/.

EU. (2017). *Proposal for a regulation concerning the respect for private life and the protection of personal data in electronic communications and repealing directive 2002/58/EC.* https://digital-strategy.ec.europa.eu/en/library/proposal-regula tion-privacy-and-electronic-communications.

EU. (2018a). *Directive 2018/1972/EC on European electronic communications code. European Parliament and the Council.* http://data.europa.eu/eli/dir/ 2018/822/oj.

EU. (2018b). *Strategic spectrum roadmap towards 5G for Europe: RSPG second opinion on 5G networks.* Radio Spectrum Policy Group (RSPG), RSPG18-005.

EU. (2019a). *Open data directive.* https://eur-lex.europa.eu/legal-content/EN/ TXT/?uri=CELEX%3A32019L1024.

EU. (2019b). *Cybersecurity act.* https://eur-lex.europa.eu/legal-content/EN/ ALL/?uri=CELEX%3A32019R0881.

EU. (2020a). *Digital markets act.* https://eur-lex.europa.eu/legal-content/en/ TXT/?uri=COM:2020:842:FIN.

EU. (2020b). *Data governance act.* https://eur-lex.europa.eu/legal-content/EN/ TXT/?uri=celex%3A52020PC0767.

EU. (2020c). *Proposal for a regulation on contestable and fair markets in the digital sector.* COM/2020/842 final.

EU. (2021a). *Artificial intelligence act.* https://eur-lex.europa.eu/legal-content/ EN/TXT/?uri=celex%3A52021PC0206.

EU. (2021b). *2030 Digital compass.* https://eufordigital.eu/wp-content/upl oads/2021/03/2030-Digital-Compass-the-European-way-for-the-Digital-Decade.pdf.

EU. (2021c). *Europe's digital decade: Commission sets the course towards a digitally empowered Europe by 2030.* https://ec.europa.eu/commission/presscorner/detail/en/ip_21_983.

EU. (2022a). *Data act.* https://eur-lex.europa.eu/legal-content/EN/TXT/?uri=CELEX%3A52022PC0068.

EU. (2022b). *Digital services act.* https://eur-lex.europa.eu/legal-content/EN/TXT/?uri=celex%3A32022R2065.

EU. (2022c). *Press release on new EU cybersecurity strategy.* https://ec.europa.eu/commission/presscorner/detail/en/IP_20_2391.

EU. (2022d). *Proposal for a regulation on harmonized rules on fair access to and use of data of the European Parliament and of the Council.* COM/2022/68 final.

E. U. New. (2020). *Cybersecurity Strategy and new rules to make physical and digital critical entities more resilient.* European Commission, 16.

European 5G Observatory. (2021). *National 5G spectrum assignment.* https://5gobservatory.eu/5g-spectrum/national5g-spectrum-assignment/.

European People's Party. (2022). *End the ever-increasing dominance of big tech companies.* https://www.eppgroup.eu/newsroom/news/end-the-ever-increasing-dominance-of-big-tech-companies.

Feasey, R. (2015). Confusion, denial and anger: The response of the telecommunications industry to the challenge of the Internet. *Telecommunications Policy, 39*(6), 444–449.

Krippendorff, K. (2004). Reliability in content analysis: Some common misconceptions and recommendations. *Human Communication Research, 30*(3), 411–433.

Kwak, K., & Yoon, H. D. (2020). Unpacking transnational industry legitimacy dynamics, windows of opportunity, and latecomers' catch-up in complex product systems. *Research Policy, 49*(4), 103954.

Lanzolla, G., Lorenz, A., Miron-Spektor, E., Schilling, M., Solinas, G., & Tucci, C. L. (2020). Digital transformation: What is new if anything? Emerging patterns and management research. *Academy of Management Discoveries, 6*(3), 341–350.

Marano, V., Tallman, S., & Teegen, H. J. (2020). The liability of disruption. *Global Strategy Journal, 10*(1), 174–209.

Massa, L., Tucci, C. L., & Afuah, A. (2017). A critical assessment of business model research. *Academy of Management Annals, 11*(1), 73–104.

Matinmikko-Blue, M., Yrjölä, S., & Ahokangas, P. (2021). Moving from 5G in verticals to sustainable 6G: Business, regulatory and technical research prospects. In *Cognitive radio-oriented wireless networks: 15th EAI international conference, CrownCom 2020, Rome, Italy, November 25–26, 2020, Proceedings 15* (pp. 176–191). Springer International Publishing.

Matinmikko-Blue, M., Yrjölä, S., Seppänen, V., Ahokangas, P., Hämmäinen, H., & Latva-Aho, M. (2019). Analysis of spectrum valuation elements for local 5G networks: Case study of 3.5-GHz band. *IEEE Transactions on Cognitive Communications and Networking, 5*(3), 741–753.

Matinmikko, M., Latva-aho, M., Ahokangas, P., & Seppänen, V. (2018). On regulations for 5G: Micro licensing for locally operated networks. *Telecommunications Policy, 42*(8), 622–635.

Matinmikko, M., Latva-Aho, M., Ahokangas, P., Yrjölä, S., & Koivumäki, T. (2017). Micro operators to boost local service delivery in 5G. *Wireless Personal Communications, 95*(1), 69–82.

McGrath, R. G. (2010). Business models: A discovery driven approach. *Long Range Planning, 43*(2–3), 247–261.

Palazzo, G., & Scherer, A. G. (2006). Corporate legitimacy as deliberation: A communicative framework. *Journal of Business Ethics, 66*, 71–88.

Phillips, M. A., & Ritala, P. (2019). A complex adaptive systems agenda for ecosystem research methodology. *Technological Forecasting and Social Change, 148*, 119739.

Seretschy, G. (2021). *The regulatory journey from a European perspective.* https://www.serentschy.com/the-regulatory-journey-from-a-european-perspective/.

Suchman, M. C. (1995). Managing legitimacy: Strategic and institutional approaches. *Academy of Management Review, 20*(3), 571–610.

Suddaby, R., Bitektine, A., & Haack, P. (2017). Legitimacy. *Academy of Management Annals, 11*(1), 451–478.

Thomas, L. D., & Ritala, P. (2022). Ecosystem legitimacy emergence: A collective action view. *Journal of Management, 48*(3), 515–541.

Turcan, R. V., & Fraser, N. M. (2016). An ethnographic study of new venture and new sector legitimation: Evidence from Moldova. *International Journal of Emerging Markets, 11*(1), 72–88.

Van Duijvenvoorde, G. P. (2020). Towards implementation of the European Union telecom code: Ex Ante reflections. *Computer and Telecommunications Law Review, 26*(7), 205–215.

Yrjölä, S. I., Ahokangas, P., & Matinmikko-Blue, M. (2021). Platform-based ecosystemic business models in future mobile operator business. *Journal of Business Models, 9*(4), 67–93.

Yrjölä, S. S., Ahokangas, P., & Matinmikko-Blue, M. (2022). Value creation and capture from technology innovation in the 6G Era. *IEEE Access, 10*, 16299–16319.

Yrjölä, S., Ahokangas, P., Matinmikko-Blue, M., Jurva, R., Kant, V., Karppinen, P., Zepernick, H. J., et al. (2020). *White paper on business of 6G.* arXiv preprint arXiv:2005.06400.

Zott, C., Amit, R., & Massa, L. (2011). The business model: Recent developments and future research. *Journal of Management, 37*(4), 1019–1042.

Zhang, N., Cheng, N., Gamage, A. T., Zhang, K., Mark, J. W., & Shen, X. (2015). Cloud assisted HetNets toward 5G wireless networks. *IEEE Communications Magazine, 53*(6), 59–65.

Weber, A., & Scuka, D. (2016). Operators at crossroads: Market protection or innovation? *Telecommunications Policy, 40*(4), 368–377.

9

Toward Anticipatory Regulation and Beyond

Georg Serentschy, Paul Timmers, and Marja Matinmikko-Blue

> The difficulty lies not so much in developing new ideas as in escaping from old ones.
>
> (John Maynard Keynes)

Advocacy for a New Policy and Regulatory Approach for 6G

Why is this chapter important and why should one read it? The sequential numbering of the various generations of mobile communications technologies (most of them known as GSM or 2G, UMTS or 3G, LTE or 4G, etc.) intuitively paints a picture of linear development, with one generation following the next. With 5G, a development has already begun that is no longer linear. Ubiquitous communication,

G. Serentschy (✉)
Serentschy Advisory Services GmbH, Vienna, Austria
e-mail: your@advice-serentschy.com

© The Author(s) 2024
P. Ahokangas and A. Aagaard (eds.), *The Changing World of Mobile Communications*,
https://doi.org/10.1007/978-3-031-33191-6_9

which includes not only person-to-person but also machine-to-machine or person-to-machine and vice versa, represents a paradigm shift from interpersonal communication to ubiquitous communication between everyone, everything, everywhere. Billions of IoT devices (i.e., real everyday physical things) are communicating with us and with each other. This development forms the basis for the digitalization of society and the economy, a development that permeates all areas of life. At the same time, it is about creating the conditions for Europe to become more competitive, to innovate, and to invest in a globalized world. It is therefore obvious that the policy and regulatory framework for such a new ecosystem cannot remain the same. In this article, we explain why a significant change in the policy and legal framework, including regulatory governance, is important and what it could look like.

Prior work on the regulation of mobile communications (Garrard, 1991; Coen, 2005; Dunnewijk & Hultén, 2007; Cave et al., 2019) has summarized the historical evolution of regulations along with the technology development from state-owned monopolies to competition. Most recently, 5G has become a source of international controversy introducing national security concerns (Robles-Carrillo, 2021) and new local deployment models (Matinmikko et al., 2018). Attention on the future development of 5G has shifted to developing policy with incomplete knowledge and under conditions of uncertainty, which calls for adaptive forms of policy and market design for innovation (Bauer & Bohlin, 2022).

P. Timmers
Oxford Internet Institute, University of Oxford, Oxford, UK
e-mail: paul.timmers@iivii.eu

M. Matinmikko-Blue
Infotech Oulu Focus Institute and Centre for Wireless Communications, University of Oulu, Oulu, Finland
e-mail: marja.matinmikko@oulu.fi

Regulation: More or Less?

Regulation is and has always been contested, and the discussion surrounding regulation hovers between 'more' or 'less' of it, and whether activity A or B should be regulated. It is striking to observe that when it comes to the issue of regulating the digital industry and the telecommunications industry, there is often a remarkably black-and-white view, with two diametrically opposed camps. On the one hand, there are the advocates of blanket deregulation who want to leave everything to the market, and on the other hand, there is the regulatory orthodoxy, the almost religious-looking representatives of the group who want to immediately press every innovation into a regulatory straight-jacket without even waiting to see how an innovation develops. This is ostensibly done with justifications such as protecting consumers from the supposedly harmful consequences of this very innovation. Then, there is also a public lament that there is no 'Silicon Valley' in Europe and that start-ups very often emigrate to the US precisely because they expect better conditions there. How does all this fit together and is there perhaps a mediating forward-looking position? In other words, are we not in Europe shooting ourselves in the foot again and again with often excessive or wrongly placed regulations and at the same time complaining that innovations and new jobs are largely in other parts of the world? What if we looked at regulation from a European sovereignty and competitiveness angle? How would it appear differently?

Developments in the field of electronic communications services and networks are driven by a seemingly never-ending stream of technological advances and the resulting product and service innovations and—not to forget—hypes. A key driver of this development is the progressive evolution of electronics and software with Moore's Law as the underlying paradigm. This essentially states that the performance of electronic circuits doubles every 12–24 months.

There is a very telling anecdote—notably from a US perspective—about the impact of Moore's Law and how important it is to recognize it in a timely fashion: Qualcomm and CDMA versus TDMA. Thomas Friedmann describes the TDMA vs. CDMA story in detail in his book *Thank you for being late* with a focus on the visionary thinking of Irwin

Jacobs (Friedmann, 2016). In the early days of mobile telephony, the European Commission mandated the standard (GSM/TDMA), and the US government allowed the market to choose. By 1991, Qualcomm had persuaded a dozen or so American telecom companies to undertake large-scale tests of CDMA technology. Once again, a serious discussion about standards started in America and in 1993, the industry body CTIA accepted CDMA as an American mobile phone standard. In the words of Irwin Jacobs, the visionary person behind the CDMA standard and co-founder of Qualcomm: "*One key reason we won was that even though CDMA was more complicated to implement, people were just thinking about the capacity of chips at that moment in time. They were not considering Moore's law that would allow the technology to improve every two years and enable the greater efficiency that could be achieved through CDMA*". People say that in hockey you don't go where the puck is, you go to where the puck is going. Qualcomm went where the puck was going—according to Moore's Law. It must be noted though, that this historical digression is only made to illustrate the power behind Moore's Law, which was well exploited by the CDMA standard. The CDMA/GSM split ended, as phone companies all switched to LTE, a single, global 4G standard, which evolved to current 5G standards.

Against the backdrop of the rapid technological developments driven by Moore's Law and in view of the question posed at the beginning of this chapter, it is not a matter of 'more of the same' but of a radical institutional renewal of regulatory work. We will show later in the text how this is intended to ensure that the challenges of digitization and a forward-looking ICT policy can be dealt with effectively.

Evolution of Regulation—An International Comparison

To start with, a brief explanation of some key concepts of telecom regulation: traditional regulation consists of the application of competition law, which by its nature is an 'ex-post' regulation, i.e., the authorities intervene after there has been misconduct by a market participant. At the start of the liberalization of the telecommunications market, it was clear from the outset that breaking up a monopoly through ex-post regulation would not be effective, since the monopolist would always be able to push the new entrant out of the market, and any ex-post regulatory intervention would be too late. This was the birth of ex-ante regulation, i.e., a set of up-front obligations was imposed on the former monopolist (in regulatory jargon, the 'incumbent') to allow the new entrant to thrive.

Sector-specific regulation of individual—so-called 'relevant'—markets through ex-ante measures as a complement to the application of competition law, which by its nature is applied ex-post, has a long tradition, especially in connection with the opening of markets that were originally monopolistic. The inherent problem with regulation, however, is finding the right balance between too much or too little regulation. However, it should not be forgotten that this is not a matter of creating a static equilibrium, but of striving for balance in a highly dynamic market that is constantly changing due to an almost endless stream of innovations. As we pointed out in an earlier article, too much regulation can impede the innovation process, while too little regulation can reduce consumer choice and/or create consumer harm.

The concept of a regulatory authority that is independent (from government and other political entities) and organizationally as autonomous as possible is a political leitmotif in Europe, North America, and some other parts of the world. However, this concept is alien to countries such as China, Japan, and South Korea, which rely on centralized government control of ICT policy and regulation. In the countries of the Middle East, there are organizationally independent regulatory authorities, but the degree of political independence is much less pronounced than in Europe and North America.

While writing this chapter, we have found that there is very little international overview of the various regulatory approaches in key industrial regions available, such as North America, Asia, and Europe. In the search for 'best practices', it therefore makes sense to make this comparison from the perspective of a 6G ecosystem. In this vein, this section describes in more detail the European regulatory journey, the regulatory rivalry between Europe and the US and some telling examples from Asia in which ICT policy and regulation are directly and exclusively in the hands of the state (China, South Korea, and Japan) as well as countries which have independent regulatory authorities, but which are under the operational control of the respective government (Taiwan, Singapore).

A standard reference for regulatory archetypes is a periodical publication provided by the ITU, the *Global ICT Regulatory Outlook*. According to the system set out by the ITU, the conceptual framework of the generations of regulation can best be described as a regulatory ladder. Generations (G) 1 through 4 reflect the evolving levels of regulatory maturity, focusing narrowly on the telecom and then the ICT sector while Generation 5 marks a major shift from sector-specific to holistic, cross-sector policies and regulations. In just over a decade, G4 has become the established standard for ICT integrated regulation led by social and economic goals. Yet, 40 percent of countries globally languish in G1 and G2, missing out on development opportunities and global digitization. G5—or the fifth generation of collaborative regulation— is the next frontier in terms of holistic digital regulation. Generations 1 through 4 are measured through the ICT Regulatory Tracker. Generation 5 is measured through the G5 Benchmark.

Regulation from 1.0 to 4.0—The European Journey

Traditional competition law, which by its very nature is based on an ex-post analysis of markets and market participants, was supplemented about 30 years ago by an ex-ante regime that imposes ex-ante obligations based on ex-post status analyses (e.g., determination of market power). However, there is *no internationally unified taxonomy for the designation of different generations of regulation*. We use the designations 'Regulation 1.0 - 4.0' in continuation of our publication *The Virtuous Circle—New Regulations, Innovation and Investment* on how to bring Europe back to the top (Serentschy, 2013). The ITU uses a different designation for 'regulatory archetypes' G1-G5 in its periodical publication *The Global ICT Regulatory Outlook*. It is important not to confuse this numbering with different generations of mobile communications technology.

This is the basis of the model of traditional telecom regulation (Regulation 1.0), which became the standard with the onset of market liberalization. The initially successful Regulation 1.0 (1990–2003) transformed inefficient monopolies in nearly all European countries into a vibrant competitive landscape. In 2003, the European institutions (the Commission, the Council, and the Parliament) embarked upon a new regime, the framework of which has survived with relatively minor modifications for the following 15 years. This landscape was characterized by telecom companies competing and the outcome was understandably hailed as a success.

However, already toward the end of this phase, voices were raised to move back to more ex-post and less ex-ante regulation. According to people familiar with the matter, this was also reflected in (to put it nicely) 'very engaged' internal debates on the European Commission level between the ex-post camp (the competition DG) and the ex-ante camp (the ICT DG). At that time, there were even considerations to reduce or abandon ex-ante regulation, which, as is well known, was only ever planned as a market opening instrument for a limited period. It is also not surprising that institutional rivalries between the ex-post camp and the ex-ante camp manifested themselves not only at the European level, but also at the national level. Interestingly, since then, the regulatory regime from 1.0 to 4.0 has always hovered between the poles of more ex-ante and less ex-post to more ex-post and less ex-ante regulation. Regulation 4.0 aiming at regulation platforms proposes again a more ex-ante approach.

This up and down movement between more ex-ante and less ex-post to more ex-post and less ex-ante regulation in the traditional telecom regulation can also be seen in the number of relevant markets, both at the retail and wholesale levels, recommended by the European Commission. This number of relevant markets reached an all-time high of eighteen relevant markets—including one broadcasting market[1]—in 2003 from originally (1997) four regulatory fields including fixed, mobile, leased lines and interconnection, only to drop to seven markets, including one retail market and 6 wholesale markets in 2007. This trend continued

---

[1] https://eur-lex.europa.eu/legal-content/EN/TXT/PDF/?uri=CELEX:32003H0311&from=EN.

with the 2014 market recommendation, which included only five whole-sale markets and no retail market. Most recently, in December 2020, the market recommendation only provides for two wholesale markets.

The current framework for telecom regulation is the *European Electronic Communications Code (EECC)*,[2] which came into force in December 2018, and which had to be transposed into national law in all member states of the Union in December 2020. However, this process is not yet completed in some member states. A comprehensive handbook for practitioners has recently been published on the legal issues associated with the application of the new regulatory framework (Liberatore & Konidaris, 2021).

Rival Regulatory Regimes: The US vs Europe

We can observe a rivalry of regulatory approaches between the US and Europe that has existed for two to three decades. While the US has relied more on laissez faire and ex-post control, Europe has increasingly become a 'world power of regulation' where far-reaching regulations and stricter consumer protection have been conceived and implemented, which have achieved international impact due to the 'Brussels Effect' (Bradford, 2020). But will it be enough for Europe to successfully defend its prosperity and innovative strength on this basis in the future? Very unlikely. To put it in the words of Cedric. O, the former French Secretary of State for Digital Transformation and Electronic Communications: "*The USA has the FAANG,*[3] *China has the BATX,*[4] *Europe can't only have the GDPR. It's time to have our own technological sovereignty and stop depending on US or Chinese solutions!*"[5] For three years, the EC has just been rolling over with a tsunami of new regulations and we have not only GDPR, but by now also many others. This *digital policy activism* includes the production of both *non-legislative* (i.e., strategies, action plans, etc.) and

---

[2] https://eur-lex.europa.eu/legal-content/EN/TXT/PDF/?uri=CELEX:32018L1972.
[3] Facebook, Amazon, Apple, Netflix, and Alphabet.
[4] Baidu, Alibaba, Tencent, Xiaomi.
[5] https://twitter.com/cedric_o/status/1336752844624551939.

legislative acts (i.e., already applicable regulations and directives, as well as proposals for such regulations or directives).

In its own words, the European Commission is determined to strengthen Europe's digital sovereignty and set standards rather than follow those of others—with a clear focus on data, technology, and infrastructure. The plan 'A Europe Fit for the Digital Age' is one of the most important and includes 16 flagship initiatives (listed in no particular order): the Artificial Intelligence Act, Data Strategy, Industrial Strategy, Chips Act, Digital Markets Act (DMA), Digital Services Act (DSA), Digital Identity (eID),[6] High Performing Computing, Digital Skills, Cybersecurity, Space (including a genuine European LEO satellite constellation), Connectivity, Contributing to European Defense, EU-US Trade and Technology Council, Cloud Strategy, the Quantum Technologies Flagship (which includes quantum computing, quantum communication, and others).

With these activities, the European Commission is, on the one hand, creating a very complex policy framework and, on the other hand, Europe has increasingly become a *global regulatory superpower* with far-reaching regulations and stricter consumer protection designed and implemented. These regulations have achieved international reach due to the *Brussels Effect* and their extraterritorial nature (Bradford, 2020). This raises immediate questions: (1) How will the enforcement of such a complex regulatory framework be organized at the national and European levels, and (2) will this plethora of regulation be sufficient for Europe to successfully defend its prosperity and promote its innovative and industrial strength on this basis in the future?

Against this backdrop, it becomes clear why we should reflect on whether we still need telecom regulation—or more regulation—in the traditional sense at all or, at best, need a radically new—*anticipatory*—regulatory approach and policies aimed at fostering strategic autonomy, digital sovereignty, and competitiveness. There are growing voices from various directions criticizing that an overabundance of regulations could

---

[6] **eID** is a set of services provided by the European Commission to enable the mutual recognition of national electronic identification schemes (eID) across borders. It allows European citizens to use their national eIDs when accessing online services from other European countries.

lead to a lack of coherent high-level vision and widen the gap between reality and political-economic ambitions.

The US vs. Europe

A current and very interesting comparison between regulation in the US and Europe can be drawn around net neutrality. While the defenders of net neutrality in Europe defend its supposed blessings in a mantra-like fashion and at the same time the ECJ's (European Court of Justice) ruling on the ban of zero-rating must be implemented by the authorities—often with noticeable reluctance—because it does not result in any advantages for consumers, in the US a phase without net neutrality regulation has apparently passed without any noticeable harm to consumers. It can be evidenced that most of the innovation still takes place in the US and not in Europe, and when the innovations are made in Europe, after a while a large part of these companies emigrate to the US or to other regions that offer better conditions. In the context of strict regulation in Europe—as in the case of net neutrality—one must ask whether this is a facilitating or hindering factor for innovation in Europe. Net (or network) neutrality is the principle that Internet and telecommunications service providers (ISPs) must treat all Internet communications equally, offering users and online content providers consistent rates irrespective of the content, website, platform, application, type of equipment, source address, a destination address, or method of communication without price discrimination. There is a lot of evidence that the hindering factors prevail in Europe because we seem to have been regulating detached from the context of innovation and industrial policy for a long time.

As an example, in 2022 Ofcom, the UK regulator, undertook a review of the UK's net neutrality framework to ensure that as technology evolves, net neutrality continues to support innovation, investment, and growth. In its newly published guidance, the regulator sets out its assessment of the issues raised and revised guidance on how the rules should apply. It looked at specific areas where greater clarity is needed to enable ISPs to innovate and manage their networks more efficiently, and to improve consumer outcomes. It also proposes to clarify its approach to enforcement where there is clear public benefit including enabling ISPs to prioritize and zero-rate access to emergency services, offer parental controls, and manage Internet traffic on airplanes and trains where there is limited capacity available. On the latter point, Ofcom states it is "*unlikely to prioritize enforcement action against traffic management on Wi-Fi services provided on board airplanes*" (OFCOM, 2023, p. 120). In addition, it says that fair use policies are compatible with the UK net neutrality rules. Ofcom has always been a European pioneer in regulation, and it is hoped, that regulators in the EU will follow Ofcom's lead in the interest of consumers, airline passengers, and innovations.

In conclusion, it seems that digital policy activism may be based on too much reliance on the broad impact of regulation and too little focus on business models, innovation policy, and investment. Regulation is only one part of the required policy measures; the other part is still incomplete. Without doubt, targeted regulation in certain areas, such as AI, is necessary to prevent harm to society and individuals. Overall, based on our analysis and experience, we believe that coherent and clear policies, combined with anticipatory and flexible regulation and the avoidance of prescriptive micro-regulation, should be core elements of a future regulatory policy.

Illustrative Examples for Regulatory Systems from Asia

South Korea

The regulatory and competition framework in South Korea differs significantly from the situation in the US or Europe. The specific situation for regulation and competition law in the country needs to be seen in the context of the country's history after the Korean war, its culture and specific industrial structure with its large conglomerates (the *chaebol*). In the words of an article, published by Moohyung Cho and Tim Büthe, South Korea, was a non-participant in the international competition regime until the 1980s, but then in the 1990s developed substantial regulatory capacity and capability. At the same time, however, to a certain extent, its policy preferences converged upon the norms and practices established by the United States and the European Union, albeit with some distinct elements. Under these conditions, the authors expect a transition from rule-taker to rule-promoter. The *Korea Fair Trade Commission* (KFTC) plays the central role as a rule-promoter. Over the course of the four decades since it first adopted the *Monopoly Regulation and Fair-Trade Act* (MRFTA), South Korea has, in the realm of competition law and policy, developed strong regulatory capacity and capability, while South Korean preferences, though distinctive on some details, have become closely aligned with the established powers' preferences.

There is also additional specific literature available, which is helpful to understand how the regulatory and competition framework in South Korea works.[7,8,9,10,11]

Japan

The first comprehensive guide to Japanese telecommunications policies in English can be found in a book by Hitoshi Mitomo (2020). This book allows readers to gain an in-depth understanding of Japanese telecommunications policies and discusses how Japan has addressed a variety of policy challenges leading to the promotion of cutting-edge technology.

The key telecommunications and ICT policy player in Japan is the Ministry of Internal Affairs and Communications (MIC), which publishes its *White Paper on Information and Communications in Japan*[12] every year. This publication is a rich source describing all policy areas, the

---

[7] https://www.kas.de/documents/262738/0/21-09-20+5G+in+Korea+Story%2C+Challenges+and+Vision+of+a+First+Mover+%28210608951%29.pdf/6efe11bb-ba66-b1b8-2c5e-3308da46c610?version=1.0&t=1631862830349.

[8] https://openknowledge.worldbank.org/handle/10986/35780.

[9] https://ijaseit.insightsociety.org/index.php?option=com_content&view=article&id=9&Itemid=1&article_id=13001.

[10] https://www.sciencedirect.com/science/article/abs/pii/S0308596117301210.

[11] https://www.sciencedirect.com/science/article/abs/pii/S0308596121001944.

[12] https://www.soumu.go.jp/johotsusintokei/whitepaper/eng/WP2021/2021-index.html.

present status and challenges for digitalization, the role of the COVID-19 pandemic for the digitalization, basic data on the ICT field and ICT policy directions.

The People's Republic of China (PRC)

A recent summary, published by Yu-li Liu and Guosong Shao in a collection of essays *Europe's Future Connected Policies and Challenges for 5G Networks*[13] and *China's 5G Development Strategies and Challenges in the Context of Global Competition,* points out that China is at the forefront of the global race in 5G and even 6G industries, and discussing its advantages and challenges may provide a reference for other countries. The development of China's 5G industry is not only supported by the government, but it also has advantages in terms of development factors, industry chain, user demand, and competitive strategies:

- The rapid development of 5G in China has benefited from the Chinese government's policy support and financial subsidies. The Chinese government has formulated policies to encourage the development of 5G networks and 5G applications and has provided 5G players with subsidies for 5G patent research and development, accelerated construction, and reduced prices of 5G products.
- China's 5G industry has a world-leading number of patents, a well-developed industry structure, influence in shaping technical standards, and a talent development strategy, all of which are important factors in promoting the development of 5G in China.
- China's 5G industry chain is robust, with upstream, midstream, and downstream segments all developing rapidly.
- China has the necessary demand conditions for 5G development, with a large subscriber base and expanding B2B business and overseas markets.

---

[13] https://liberalforum.eu/wp-content/uploads/2022/06/Europes-Future-Connected_ELF-Study_Techno-Politics_vol.2-2.pdf.

- Chinese 5G operators have adopted different service strategies to address to business and to consumer business, developing differentiated value-added services to improve their competitiveness in the market.

Looking forward, China's rapid development of 5G has laid a good foundation for the development of 6G in the future. The experiences China has gained in 5G development can be used as a reference for European countries. European countries can explore their own development models, tailored to their national conditions and characteristics. An analysis of the opportunities and challenges of 5G development in China may help to facilitate the development of 5G and 6G in Europe.

A detailed description of the policymaking bodies and regulatory authorities in the PRC can be found in an article by Shihui Partners,[14] a private Chinese law firm. Raymond Wang, the author of this article, describes the situation in detail: "*Over recent years, China has continuously liberalized the access restrictions for TMT through various policies. At the same time, China is also strengthening its supervisory capabilities in the fields of network and data security. With the formal entry into force and gradual implementation of a series of laws and regulations, such as the Cybersecurity Law, Data Security Law and the Personal Information Protection Law, regulatory enforcement activities, particularly in the online space, have become increasingly prominent. Moreover, it can be assumed that cybersecurity and data protection compliance issues (e.g., those relating to listed companies) will become an increasing concern for regulators*".

Meanwhile, various incentive measures and preferential policies have been designed to ensure the protection and conditions for the expansion of fair competition and the development of a healthy market. The TMT regulation in China divides all telecommunications into two categories:

- Basic telecommunications services (BTS) and

---

[14] https://thelawreviews.co.uk/title/the-technology-media-and-telecommunications-review/china.

• Value-added telecommunications services (VATS).

BTS essentially refer to the provision of infrastructure facilities and basic voice and data transmissions, both domestically and internationally, while VATS refer to the provision of specialized services via the basic infrastructure facilities. China adopts a strict licensing system for the telecoms industry, and telecoms operators are required to obtain a license to engage in either BTS or VATS. To fulfill its commitments to the World Trade Organization, China is gradually opening its telecoms industry to foreign investment. Among all the VATS, Internet content services and e-commerce have grown at a rapid pace in recent years. Following the prosperity of the Internet industry, online IP infringement, unfair competition, and anti-counterfeiting, antitrust, cybersecurity and data security risks, and personal information (PI) protection are issues that are starting to become of greater concern to telecoms operators.

Taiwan

Since its establishment on February 22, 2006, the National Communications Commission (NCC) has been the authority responsible for regulating telecommunications and broadcasting services in Taiwan. Originally, this authority belonged to both the Directorate General of Telecommunications and the Department of Broadcasting Affairs of the Government Information office; the merged mandate of the NCC is a milestone which is indicative of the advent of digital convergence.

Under the present trend, the convergence of telecommunications, broadcasting, and information networks has become inevitable. As a result, a regulatory agency must govern the communications sector with a broader and more accurate strategic insight, as well as an open and more efficient administration. It also must coordinate the efforts of the executive and legislative branches, as well as those of the private sector to respond to the rapid development, the expectations of the public, and the transformation of society.

The four policy goals of the NCC are: (1) promote the sound development of communications; (2) safeguard the rights of the people; (3) protect the interests of consumers; and (4) raise multicultural diversity.

These four policy goals constitute the administrative principles of NCC. In addition, the four administrative beliefs of the NCC include independence, responsibility, balance, and maximum benefits for the public. The NCC carries out the four policy goals in accordance with its administrative beliefs. Together, they make up the administrative framework of NCC.

The NCC is the first legitimate regulatory agency in Taiwan independent from an executive branch. The NCC analyzes the development of digital convergence to formulate a direction for communications regulatory reform in accordance with the basic supervisory principles of the Fundamental Communications Act as well as national policies and objectives. The NCC aims to regulate the communications sector from an objective, neutral, and professional standpoint, to ensure effective competition in the market, safeguard public interest, promote the development of communications services, and thereby enhance the nation's competitiveness. More details about telecom regulation in Taiwan can be found online.[15]

Singapore

The Infocomm Media Development Authority (IMDA) is in charge for telecom regulation in Singapore. As the name suggests, it is a convergent telecommunications and media regulator with the following self-description: "*IMDA develops and regulates the converging infocomm and media sectors in a holistic way, creating a dynamic and exciting sector filled with opportunities for growth, through an emphasis on talent, research, innovation, and enterprise. As a statutory board in the Singapore government, it seeks to deepen regulatory capabilities for a converged infocomm media sector, safeguarding the interests of consumers and fostering pro-enterprise regulations. With more pervasive use of data, IMDA will also continue to promote and regulate data protection in Singapore through the Personal Data Protection Commission, which will be part of the IMDA. This will ensure that public confidence in the private sector's use of personal data is safeguarded*".

---

[15] https://www.lexology.com/library/detail.aspx?g=d3aa1226-9fd7-4594-9bad-3e801630bc66.

In parallel to the United Kingdom, Singapore also has extensive activities and experience with *regulatory sandboxing*. The IMDA operates a program called *Policy Prototyping* with the goal of promoting the responsible use of data and supporting data-driven innovation by businesses, as well as building consumer confidence. To this end, IMDA works closely with and supports industry through policy prototyping and data regulatory sandboxing. Similarly, cross-sectoral activities are operated by the Monetary Authority of Singapore (MAS), the Energy Market Authority (EMA), and the National Environmental Agency Regulatory Sandbox.

Against this backdrop, it is not surprising that Singapore is one of the world's fifth generation (G5) leaders in terms of regulation, according to the International Telecommunication Union (ITU) rankings. The ITU describes the country as a digital pioneer with a mature ICT framework. Singapore has succeeded in transforming the country's industry and government institutions to leverage technology for the benefit of digitalization, and has done so through cross-sector, collaborative regulation. Specifically, in collaboration, Singapore leads the ITU G5 benchmark (Table 9.1), along with countries such as Norway and the United Kingdom (ITU, 2020).

Anticipatory Regulation and Beyond

In this sub-chapter, we present a *proposal for a radically renewed regulatory approach in a 6G ecosystem*. This sub-chapter is structured in the following way:

- An advocacy for a wide-ranging digital policy authority for a 6G ecosystem.
- It is time to act now—the 'Burning Platform'.
- Regulatory governance, institutional reform, and a new mandate for regulators.
- A step-by-step approach.
- Conclusion and areas for future research.

Table 9.1 5th Generation Regulation (G5) countries by score, rank, and the ICT regulatory tracker

G5 Benchmark 2019			ICT Regulatory tracker	
Country	Score	Rank	Tracker rank	Gen
1. Norway	39	1	3	G4
2. Singapore	39	1	26	G4
3. Japan	37	2	106	G3
4. Estonia	37	2	47	G4
5. United Kingdom	37	2	4	G4
6. Canada	37	2	58	G4
7. Kenya	37	2	45	G4
8. Croatia	36	3	11	G4
9. Romania	36	3	23	G4
10. Spain	36	3	52	G4
11. Germany	36	3	16	G4
12. Albania	35	4	69	G4
13. Brazil	35	4	36	G4
14. Netherlands	35	4	19	G4
15. Sweden	35	4	33	G3
16. Morocco	35	4	36	G4

Source ITU[16]

Against the backdrop of the strong role of large digital platforms in the provision of interpersonal communications services, it is no exaggeration to say that traditional telecommunications companies will increasingly lose, or have already lost, their original key and universal role in the electronic communications market. Interpersonal communication has already become the GAFA's (Google, Apple, Facebook, Amazon) home-turf, and these companies are also capturing a growing chunk of the B2B services. Whether one likes this result or not, the fact remains that most telecommunications companies in Europe have so far not—or only incompletely—succeeded in establishing additional services beyond the provision of connectivity successfully on the market on a sustained basis under the current conditions (especially under the pressure of OTTs—over-the-top players—and regulation). In addition, many telecom companies are not managed according to the principles of

---

[16] https://www.itu.int/en/ITU-D/Conferences/GSR/2019/Documents/G5-Benchmark_atGSR19.pdf.

long-term infrastructure investment but are managed under the pressure of short-term financial results and capital market expectations.

Focus on Connectivity

Perhaps under these conditions, it would be worth considering as a telecommunications company to focus on creating ubiquitous connectivity with customized service levels. Surely that would not be a bad thing! However, there is not much time for this, because the OTTs themselves are striving to provide connectivity. The discussions about the introduction of openRAN on a large scale have, in particular, made it clear to all stakeholders that it is necessary to build up more technical capacities in the telecom companies in order to carry out the system integration of the components themselves, for example.

Mobile telecommunications have become a layer in a towering stack of technologies and applications. Downwards this includes devices (IoT, wearable, etc.) of all sorts that are becoming smart and may run substantial data analytics themselves and have direct and private network connections to edge cloud. Upwards this includes dataspaces, cloud, content delivery, services for trust, and security and access management, and apps of all sorts for communication purposes.

On the one hand, commercially, telecoms are squeezed between actors from the lower and the higher layers of the stack. OTTs are delivering communication services 'on-top' of the connectivity provided by traditional telecom companies, but the regulatory pressure is not evenly distributed between the players in the stack. From this perspective, 6G is an enriched connectivity and communications layer in an ever-richer technology stack. Telecoms may be seen as being controlled ever more by other players above or below telcos in the stack. On the other hand, telecoms seek to enhance value and bring these other layers closer to their core business. From this point of view, future telecoms regulation, or even a reduced regulatory environment, should serve to provide telecoms with that room to grow and compete. Innovation and industrial capabilities are the building blocks for the regulatory environment of a wider 6G ecosystem. From this perspective, the richness of 6G allows it to break out of the box. 6G would be a commercial gamechanger, perhaps even bringing a new paradigm for business models.

While we can see these two economic-technical futures, we should not forget that telecoms and their regulatory legacy also come from a history of serving essential public interests for communication and connectivity. There is a path-dependency that we need to consider. Looking forward to 6G, it is the right time to again ask the fundamental questions: what is the role for regulation of 6G telecoms in view of economic, social, and democratic interests, and what type of regulation, if any should there be? Fig. 9.1 illustrates this situation, upwards directed arrows indicate in a qualitative manner of "growing importance/relevance", and downwards vice versa.

We can start by analyzing what it will take to realize the future that today seems most challenging, namely: how to enable the second future and contain or avoid the risks of the first potential future. To do so, we draw up a timeline that shows business activities and threats/risks to these, and the role of regulation. From this we can then assess the role of regulation vs each of the interests and the moment of intervention. The latter makes anticipatory regulation 'live'. To make a connection with the chapter that discusses sovereignty in this book, one way would be to look

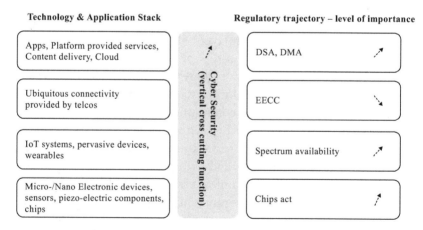

Fig. 9.1 Regulatory trajectories from traditional to anticipatory regulation

at anticipatory regulation from the perspective of strategic autonomy, digital sovereignty, and competitiveness as we want it to be.

Technological Development and Regulation

Against the backdrop of the rapid technological developments driven by Moore's Law, in our view it is not a matter of 'more of the same' but of a radical institutional renewal of regulatory work. In the following sub-section, we will show how this is intended to ensure that the challenges of digitization and a forward-looking ICT policy can be dealt with effectively. The traditional approach taken by representatives from the regulatory orthodoxy (most policymakers and many regulators) is characterized by a policy that is often isolated and disconnected from markets and innovation. It is characterized by an intrinsic up and down of the number of relevant markets and the hovering between more ex-ante and ex-post regulation.

In the following sub-sections, we attempt to answer the question of how to get policymakers out of the box. Ultimately, a combination of strong leadership at the top and experts with broad multidisciplinary backgrounds seems to be a possible recipe for success.

Europe lacks a smart combination of regulation and industry/ innovation policy (Serentschy, 2021) as a 'third way' between the poles of more or less ex-ante regulation and ex-post regulation. This approach would also take the geopolitical focus into account, as argued in the sovereignty-related chapter of this book. How such a combination of regulation with industrial and innovation policy can be organized and managed will be described in the following sub-sections. Europe as a regulatory superpower, combined with the Brussels Effect, behaves like a referee, but the referee does not make the game! We need to get much more on the playing field in Europe! Europe's digital policy approach is too focused on regulation. The idea of gaining industrial competitiveness and innovation power by regulating others is like stopping halfway. To make our position clear: We are not advocating for a blanket reduction or abolition of traditional telecom regulation, which is being cut back all the time anyway. In our view, what is crucial is how a smart combination of regulation with innovation and industrial policy can be achieved and what the necessary political pre-conditions and institutional structures are for this to happen.

A Call for a Wide-Ranging Digital Policy Authority

Regulation in a broader sense must provide the policy, operational and competition framework for a particular sector. One of the challenges in a very fast-moving and innovation-driven market is how to ensure future-proof and targeted regulation in a rapidly changing environment, driven by dynamic technological advancements and market developments. The characteristics of these changes include, on the one hand, the declining relevance of traditional telecommunications services and, at the same time, the increasing narrowing of telecommunications companies' business activities to the provision of connectivity, and, on the other hand, the growing importance of digital platforms for the delivery of communications services. Against this backdrop, the strategic question arises as to what future-proof and targeted regulation of the ICT sector will look like and what will be required to promote innovation and the industrial ecosystem in Europe. Another aspect which should be kept in mind in this context is that the specifics of 5G development should be considered for an effective 6G policy framework.

The specifics of 5G development include, for example: (1) concentration of relevant intellectual property rights—IPRs (i.e., patents, etc.) with a few (corporate) players; (2) greatly varying efficiency in the use/exploitation of IPRs by those very players (e.g., Huawei has wide-ranging IPRs, but can only use them to a limited extent due to US restrictions); and (3) the increasing complexity of standardization. In some areas, smaller specialized players are driving the development, initially working proprietarily in mobile private networks (MPNs). In the area of individual components, such as core networks, there are many providers, each of which focuses on only one or a few topics or components. These players try to bring their innovative approaches into the standardization processes to broaden their business perspectives and market. At the same time, (4) a fragmentation of the technological development can be observed in technologies such as LoRaWAN (Long Range Wide Area Networks), which may weaken standardization ambitions.

This is a call to action for all relevant ICT policy bodies (ministries, regulators, general competition authorities, R&D, and innovation institutions, etc.) to develop an effective policy and institutional framework to improve the conditions for promoting the ICT sector. Clearly, the introduction of policy, regulatory, and institutional innovations also require challenging and time-consuming debates with the political and regulatory orthodoxy about the need for change. A promising approach to a forward-looking debate might be to paint a *burning platform* scenario.

The 'Burning Platform'

Providing key stakeholders with a compelling vision of the future is a typical approach to inspiring change and conveying a message that the future painted on paper will be better than the current state. The problem with this approach is that a grand vision is not always enough to get everyone mobilized and moving in the same direction. Another approach to motivating people to change is to create a 'burning platform' scenario. The concept of a burning platform comes from the analogy of standing on an oil platform at sea that is on fire—the urgency to save yourself is so great that you act and jump off. In business, a burning platform is a term used to describe the process of helping people to see the dire consequences of not changing. By sparking just enough concern about what happens if the status quo remains the same, people embrace change. The problem with this approach is that people usually do not know what lies on the other side of the change.

Against the background of the changes described previously, the question arises as to whether the existing institutional set up is still fit for purpose, since regulatory work is often carried out detached from industrial and innovation policy. On the one hand, Europe has increasingly become a 'world regulatory super-power', with far-reaching regulations and stricter consumer protection designed and implemented, which have achieved international impact due to the Brussels Effect. But will this not be enough for Europe to successfully defend its prosperity and foster its innovative and industrial strength forever, as pointed out by Cedric. O, the former French Secretary of State for Digital Transformation and Electronic Communications. In response to this dilemma, the EC is just rolling out a tsunami of legislative acts and we have not only GDPR, but by now also the DMA, DSA, Chips Act, Data Act, Artificial Intelligence Act, eID, etc. However, there is a growing awareness that an over-abundance of regulations will not guarantee better outcomes in terms of innovation strengths and competitiveness. Against this backdrop, it is clear why we need new policies aimed at fostering competitiveness, strategic autonomy, and digital sovereignty for Europe. An additional highly topical aspect for the efforts for strategic autonomy which should not be forgotten at this point is the further geopolitical threat scenario for Europe, which extends the strategic need for action in the direction of raw materials, energy supply, securing global supply chains, etc. For more details, see Timmers (2022).

Digital Dependency

A recent research report by Mayer and Lu, (2022), funded by the Konrad Adenauer Foundation concluded, that *Europe has not fully recognized its digital dependency yet*. The results of measuring digital dependence suggest a sober reassessment of the status of 'digital autonomy'. The Digital Dependency Index DDI has implications for various actors involved in digital policymaking at the national and EU level. The key message is that the *degree of digital dependence of EU members is far greater, more pervasive, and multifaceted than often assumed*:

- *European countries are falling behind in every dimension* compared to China, South Korea, and the US. In the last decade, Europe's digital autonomy has eroded as digital interactions have become more asymmetric with China (ICT trade dependence), with the US (infrastructure and platform dependence), and the East Asian region (IP dependence).

- *European governments need to rethink their entire approach to digital technologies.* If the goal of improving technological autonomy is taken seriously, a consistent implementation of strategic autonomy goals in political, temporal, and financial terms as well as with a strategic vision would be required. This consistency is missed by the authors.
- *European companies and governments* should put a stronger emphasis on reducing their growing dependency on foreign IPs in the ICT field.
- Germany [and other European countries] should *draw lessons from other technological middle powers, especially from South Korea and Japan.*

These findings among others clearly show that the current regulatory governance and institutional set up is not sufficiently supportive for achieving digital sovereignty and strategic autonomy. Considering the amazing supremacy of the US in the field of innovation, these facts should generate a sense for urgency.

Institutional Reform and a New Mandate

The question of whether telecommunications regulation could not be institutionally combined with other policy areas to exploit substantive or financial synergies has repeatedly arisen in connection with public governance issues. In the past, there have been various approaches to this, such as the multisector regulator (the German Federal Network Agency BNetzA is a prominent example), or the widely used combination of telecom regulation with (electronic) media regulation. Experience shows that these approaches are outdated and only functional to a limited extent, since there is obviously no common denominator between these sectors due to different regulatory objectives, political accountability and, moreover, the expected synergy gains could not be realized.

The conclusion from the analysis is that telecoms regulatory authorities in the EU have become ever less relevant as EU telecoms are increasingly bypassed and left behind by large digital platform and cloud providers, and China's techno-state policy. To make matters worse, regulatory authorities have focused only on market conditions rather than the wider interests of competitiveness and digital sovereignty.

The reasons for the failure are threefold: (1) not being mandated to address the wider interests; (2) not being mandated to use a wider toolbox than traditional telecoms regulatory policy; and (3) a degree of blindness to large changes which has been made worse by institutional fossilization. Telecoms regulatory authorities need to break out of their sectoral silo. Fortunately, many are willing to do so and 6G offers an opportunity. The implication is that a *smart combination of regulation, innovation policy, and industry policy* is needed to promote regulatory innovation such as *anticipatory regulation or sandboxing and agile policy-making* in the telecommunications and digital ecosystem (see also in one of our earlier articles[17]). In a different context of collaborative ecosystems and cloud applications, Berk and Saxenian (2022, p. 64) claim that "*Our research suggests that competition policy, innovation policy, and industrial policy should be seen as complementary, particularly for supporting today's collaborative ecosystems*".

Smartness refers to an agile form of organization in which project teams come together to find an adequate solution, depending on the complexity of the problem and whether the path to the solution is known. The opposite of smart would be a static, traditional form of organization dealing with the different aspects of ICT policy in vertical silos such as infrastructure, services, applications, etc. The key point here is that the fast pace of technological development requires agile action and the teams of experts working on these issues are moving between the middle complex to chaotic, and therefore need to be organized and managed accordingly.

Step-by-Step Toward a Wide-Ranging Digital Authority

To make Europe's digital policies more impactful, we are advocating for a step-by-step introduction of a wide-ranging digital policy authority for a 6G ecosystem and recommend going a step further and expanding telecommunications regulation in combination with innovation and

---

[17] https://www.serentschy.com/wp-content/uploads/2022/01/20210802_Innovation-and-Regulation-FINAL.pdf.

industrial policy as a first step at the national level in the direction of a far-reaching digital policy authority. The term 'authority' is deliberately kept ambiguous in this context: on the one hand, as an authoritative body, i.e., a group of experts, and on the other hand, a unit that makes decisions, a regulator. This ambiguity was chosen to outline a possible development path from a 'light' scenario with a group of experts advising the decision-making bodies to a potentially full scenario at a later stage with more powers, including decision-making powers.

There is a wealth of literature and relevant experience on various governance models for successful ICT policy implementation (including the failed approaches with a 'digital ministry'[18]), but not explicitly in the sense described here. Considering the efforts toward digital sovereignty and strategic autonomy for Europe, this combined entity—covering telecom, standardization, digital platforms, innovation, and industrial policy aspects—could provide expert, non-partisan advice at the political level, support the market players in implementing laws and intervene in cases of abuse, help protect consumer interests, and make an important contribution to a successful ICT policy.

There are the first promising approaches to an integrative regulatory approach in Europe. The Danish government made a conscious decision a long time ago to bundle the digital agendas in the Ministry of Finance and not to go down the route of a Digital Ministry, which Sweden once introduced and then discarded. The Ministry of Finance is generally the only one that can exercise a lateral control function with its allocation policy for financial resources. A benchmark was set in the UK, when the Competition and Markets Authority (CMA), the Information Commissioner's Office (ICO), and the Office of Communications (Ofcom) formed the Digital Regulation Cooperation Forum (DRCF) in July 2020. Building on the strong working relationships between these organizations, the forum was established to ensure a greater level of cooperation, given the unique challenges posed by regulation of online platforms.

---

[18] Does a **"Digital Ministry"** make sense? For example in the context of setting up a "digital ministry" in Germany (in German language): https://regierungsforschung.de/wp-content/upl oads/2021/09/Studie_Digitalland_Deutschland_regierungsforschung_de_NRW_SoG_Accent ure_210920-1-1.pdf.

It is important to keep in mind that such a fundamental conceptual and institutional change cannot be achieved simply by flipping a switch. If the burning platform scenario creates enough urgency to convince key stakeholders of the need for reform, a *development path* must be established *in two dimensions*. In terms of the *competence scope*, at the national level, for example, one could start with a 'light' scenario where a public think tank composed of existing entities in the fields of telecom regulation, innovation and research, and industrial policy, etc., advises the decision-making bodies, to a potentially full scenario at a later stage with more powers, including decision-making powers (i.e., regulatory powers).

In the *national vs. supra-national* dimension, one could gradually develop the institution from the national toward the European level. In this context, BEREC would also need to be further developed, e.g., by having a subgroup dealing specifically with transnational and European issues. In the longer term, under a fully digital single market scenario, the establishment of a European regulatory and digital agency appears to be without alternative. However, there is still a considerable way to go.

Intuitively, one might think that 'digital' is a global issue and therefore requires a global or supra-national body to deal with all aspects of digitization. In this sense, responsibility for this matter could be attached to a UN organization, possibly the ITU. For practical reasons, this option is left aside here. At the European Union level, the European Commission can be seen as a European digital authority. To be as specific as possible and to start with, the focus is here on the national level. This also has the side effect that important players in this field, such as the United Kingdom or Switzerland, who are not members of the EU are conceptually included in these considerations. As a first step, the concept of a far-reaching digital policy authority is understood at the national level as described here, but without focusing on a specific country. Further steps toward a European instance need to be developed.

Summary and Areas for Future Research

One of the aims of our proposal for a regulatory system in the 6G ecosystem is to help Europe become more competitive, innovate, and invest in a globalized world. However, this goal can only be achieved inadequately, or at least not efficiently, under the current conditions. Regulators largely operate in a silo, widely detached from innovation policy and industrial policy and they are lacking the mandate and the tools to act in a wider area with adjacent public entities in the research, innovation, and industrial policy fields. Our proposal aims to create a framework so that regulation, industry, and innovation policy can work together to effectively support the above goal.

The concrete operational implementation of such a new framework requires, on the one hand, an even better design of the theoretical underpinning and, on the other hand, an adapted translation into policy practice. This may vary from country to country, but in any case, requires a synchronized approach at the supra-national level (especially the EU) and the national levels.

Bibliography

Bauer, J. M., & Bohlin, E. (2022). Regulation and innovation in 5G markets. *Telecommunications Policy, 46*(4), 102260.

Berk, G., & Saxenian, A. (2022). Architectures of participation. *Issues in Science and Technology, 38*(4), 62–69.

Bradford, A. (2020). *The Brussels effect: How the European Union rules the world*. Oxford University Press.

Cave, M., Genakos, C., & Valletti, T. (2019). The European framework for regulating telecommunications: A 25-year appraisal. *Review of Industrial Organization, 55*(1), 47–62.

Coen, D. (2005). Managing the political life cycle of regulation in the UK and German telecommunication sectors. *Annals of Public and Cooperative Economics, 76*(1), 59–84.

Dunnewijk, T., & Hultén, S. (2007). A brief history of mobile communication in Europe. *Telematics and Informatics, 24*(3), 164–179.

Friedmann, T. (2016). *Thank you for being late*. Allen Lane.

Garrard, G. (1991, December). The regulation of mobile communications. *1991 sixth international conference on mobile radio and personal communications* (pp. 131–135). IET.

Introducing the new 5G benchmark [ITU, 2020]. https://www.itu.int/en/ ITUD/ Conferences/GSR/2019/Documents/G5-Benchmark_atGSR19.pdf.

Liberatore, F. & Konidaris, J. (2021). *EU Electronic Communications Code handbook*. Bloomsbury Professional.

Matinmikko, M., Latva-aho, M., Ahokangas, P., & Seppänen, V. (2018). On regulations for 5G: Micro licensing for locally operated networks. *Telecommunications Policy, 42*(8), 622–635.

Mayer, M., & Lu, Y. (2022). *Digital autonomy? Measuring the global digital dependence structure*. Center for Advanced Security, Strategic and Integration Studies. https://www.kas.de/documents/252038/16166715/Digital+Autonomy+-+Measuring+the+Global+Digital+Dependence+Structure.pdf/fb97d384-53fd-b747-908f-2c86e8d0674b?version=1.2&t=1651491803819.

Mitomo, H. (Ed.). (2020). *Telecommunications policies of Japan*. Springer.

OFCOM. (2023). *Net neutrality review*. https://www.ofcom.org.uk/__data/assets/pdf_file/0028/245926/net-neutrality-review.pdf.

Robles-Carrillo, M. (2021). European Union policy on 5G: Context, scope, and limits. *Telecommunications Policy, 45*(8), 102216.

Serentschy, G. (2013). *The virtuous circle—New regulations, innovation and investment—How to bring Europe back to the top*. https://www.serentschy.com/wp-content/uploads/2014/02/The_Virtuous_Circle_Buch-download.pdf.

Serentschy, G. (2021). *Innovation and regulation in the digital communications field*. https://www.serentschy.com/the-regulatory-journeyfrom-a-european-perspective/.

Timmers, P. (2022). *Digital industrial policy for Europe (CERRE)*. https://cerre.eu/wp-content/uploads/2022/12/Digital-Industrial-Policy-for-Europe.pdf.

10

Sovereignty and 6G

Paul Timmers and Georg Serentschy

Change takes courage.

(Alexandria Ocasio-Cortez)

Strategic Autonomy and Sovereignty

Sovereignty is a concept that is much debated among political scientists, without arriving at a single or clear definition. The origins lie in the establishment of the system of European states in 1648 in the Treaty of Westfalen, which sought to bring peace and stability after 30 years of war on the European continent. Key elements include 'the sovereign'

P. Timmers (✉)
Oxford Internet Institute, University of Oxford, Oxford, UK
e-mail: paul.timmers@iivii.eu

G. Serentschy
Serentschy Advisory Services GmbH, Vienna, Austria
e-mail: your@advice-serentschy.com

© The Author(s) 2024   **253**
P. Ahokangas and A. Aagaard (eds.), *The Changing World of Mobile Communications*,
https://doi.org/10.1007/978-3-031-33191-6_10

embodied in the state responsible to ensure security (the Leviathan of political philosopher Hobbes), non-interference between states, and recognition of borders. At the time it seemed that unavoidably this would imply an 'anarchy of states' as well as arms races between states. However, thinking has much evolved, and a rich governance of the international system of states has emerged through multilateralism, which includes a concrete manifestation such as international standardization.

Loosely speaking, sovereignty is about territory and borders, people, 'our' values, and resources that 'belong to us'. Sovereignty requires internal legitimacy of the authority toward the people. Sovereignty also requires external legitimacy, that is, recognition by other states (Biersteker, 2012; Timmers, 2022a). It may well be that an agreement on what state sovereignty is, cannot be achieved, because it is an 'essentially contested concept' (Gallie, 1956).

Likewise, strategic autonomy is not uniquely defined, but has been far less discussed by academia until recently. The term has one origin in India, after World War II. Its meaning was to be able to exercise India's wish to remain independent from Beijing, Moscow, and Washington. Another origin is, also after World War II, in French military doctrine, where its meaning concerns having the capabilities and capacities necessary in order to exercise French power wherever necessary in the world ('frappe de force' with nuclear capability and aircraft carrier, etc., [Wikipedia, 2022]). In recent years, the thinking has evolved under the influence of geopolitics, pervasive and disruptive digital developments such as the rise of dominant platform companies, and global challenges from cybercrime to pandemic and climate. France itself formulated a wider non-military industrial scope for strategic autonomy, which was further widened in 2017 at the EU level as a concern about autonomy in economy, society, and democracy (European Commission & European External Action Service, 2017). Subsequently, a plethora of interpretations were put forward. Now that the dust has settled, it is clear that what these have in common is that strategic autonomy is about the capacities, capabilities, and control—or three Cs—necessary to safeguard sovereignty. That is, strategic autonomy is the means, and sovereignty is the end (Timmers, 2022b). Capabilities are what one knows to do. Capacities are how much one can do. Digital sovereignty is by most authors interpreted as digital strategic autonomy, i.e., strategic autonomy in the digital domain. Strictly speaking, it is therefore a misnomer but the use of the terminology of digital sovereignty, technological sovereignty, and others such as health sovereignty, have become widespread.

In the military domain, the use of the term strategic autonomy is still probably the clearest, namely the capabilities, capacities, and control in the military domain to defend national security and sovereignty more or less as circumscribed above (as is as such also written in 'task description' of ministries of defense, see e.g., Ministère des Armées | Ministère des Armées [2022] and Ministry of Defense, UK [2022]).

As in subsequent sections, let's make the link to 6G. Where people talk about digital strategic autonomy and make this more specific to certain technologies such as semiconductor strategic autonomy, should we also talk of 6G strategic autonomy or 6G sovereignty? Does it make logical, political sense, economic and social sense? We will further address these questions below but for now suffice to say: be prepared for these terms to be touted, whether sensible or not!

Pathways to Strategic Autonomy

National or regional security is central to sovereignty, as is the power to decide and act upon one's own future in terms of economy, society, and democracy. One pathway to strategic autonomy could be self-sufficiency or *autarky* of the necessary capabilities and capacities. This route is perhaps possibly for the largest countries, the USA, and China, but not for the EU or individual smaller states. It would simply be too costly and essential resources are lacking such as—in the domain of telecommunications—materials for electronic components. The more realistic pathways then are:

A. *Strategic partnerships of like-minded actors*, i.e., states and industry that sufficiently trust each other and share values that respect and reinforce the sovereignty of each of the states in the partnership. Even these strategic partnerships of the like-minded may not be self-sufficient and could be threatened by third states. They will likely have to live with strategic interdependency of non-like-minded parties and establish a reasonably stable strategic equilibrium. What this means in (6G) telecoms will be illustrated shortly.

B. Promote and support *global commons*, i.e., raise the challenge of three Cs to the global level to seek collaboration under forms of governance that maintain the global common good and deter free riders (Ostrom, 2015). In the field of technology, we know this from Internet domain name management in ICANN and Internet protocol management in IETF. An interesting example because of its well-evolved governance is the SWIFT system of private and central banks for international electronic transactions, though that has its own peculiarities and deficiencies (Cowhey & Aronson, 2017).

C. Handle strategic autonomy based on *risk management*, including state-of-the-art risk assessment, and possibly based on regulation. This is how countries over many years have dealt with technologies that are critical for sovereignty, telecommunications included. In the EU, risk management is the nature of key digital legislation such as the Electronic Communications Act, the General Data Protection Regulation, and even the first version of the Network and Information Security Directive.

A state or alliance of states may pursue strategic autonomy through a combination of these pathways, dividing up the area of concern—say telecommunications—into subareas. For instance, hardware encryption intended for state secrets might only be dealt with by self-sufficiency, while core network management involves a strategic partnership with trusted partners, and global interoperability may be handled as a global common good, and the less critical and more local remainder through risk management.

The choices are highly political and taken under the pressure of forces that threaten sovereignty, i.e., geopolitical tensions, digital dominance by companies under third-country control, and serious digital disruption such as cyber-attacks and theft of critical intellectual property. Given these three options, let's consider what future 6G strategic autonomy may be based on according to past and current approaches to telecoms in 3G, 4G, and to some extent 5G and consider the EU, the USA, and China.[1]

The European Union

In the EU, telecommunications privatization and liberalization since the 1970s and the general basis in EU telecoms policy since the 2000s of open liberal market economy thinking has led to a retreating role of the state, national interests, and security. Increasing consumer benefits have been delivered through economic legislation based on free trade in the EU internal market, but strategic autonomy has not been on the agenda, and neither has there been a strong role for national security. Past EU telecoms policy, as is still *en vigueur*, has been largely strategic autonomy-agnostic, except for telecoms security, which has been risk management-based (see above). Open-to-the-world thinking has also been dominant in the approach to EU and international standardization which, moreover, has been essentially fully left to industry without much government involvement—which is a cause of the 5G security problem, see below, and (Timmers, 2020)—and in the EU's large R&D program in which, for instance, Chinese companies were also present.[2] The latter, by the way, did not come with corresponding opening-up of Chinese R&D programs. Again, all this lines up to confirm that at best strategic autonomy had been a risk management approach and even then, with weak risk management, This, however, changed from 2017

---

[1] For more insight in and detail of the telecoms regulatory landscape, see Serentschy et al. (2022).

[2] China-registered organizations participated in 91 telecoms/broadband/cloud-related projects in the EU's R&D Framework Programs (FPs), from 1998 to 2020, with a steep rise from 2014 to 2020 in Horizon 2020 or FP8. More specifically, Huawei is a large participant be it from its subsidiaries in Europe rather than from its headquarters in China. Namely in 3 projects in FP6 (2002–2006), 9 projects in FP7 (2007–2013) and 25 in Horizon 2020 or FP8 (2014–2020). Noteworthy is that direct Chinese participation slumped to only 2 projects in the first two years of current program, Horizon Europe, but that Huawei continues to have a strong presence with already 10 projects in Horizon Europe, comparable to the past. This includes 6G visioning projects such as the EU-funded SCION project (which runs until 2027 in Horizon Europe or FP9). Huawei is also a member of the 6G-Industrial Alliance which is the industry representation in the EU's 6G-Smart Networks and Services Joint Undertaking. This initiative has as its missions 'Fostering Europe's technology sovereignty in 6G' and 'Boosting 5G deployment in Europe'. Details (Table 10.2) at the end of chapter.

onwards, due to the 5G security debate triggered by concerns from the USA. Gradually, awareness grew of national security risks which also drew attention to the ever-growing market share of Huawei and ZTE and the related risk of marginalization of the European telecoms industry, notably Ericsson and Nokia. 5G security was raised to the top level and EU countries collaborated with the support of the European Commission on a 5G Security 'Toolbox' to assess technical and political risks of foreign vendors. What is quite remarkable is that, where the core of 5G security are national security concerns and where national security is explicitly excluded from the mandate of the EU (Art 41 of the Treaty on the Functioning of the European Union), countries decided to collaborate precisely at the European level on this core issue! Moreover, though it took much more time, until 2021, EU policymakers realized the weakness of presence in international standardization, notably in telecoms. A strongly worded revision of EU standardization policy was then issued, which notably singled out the EU contribution to telecoms standardization to be brought under strategic autonomy control. Shortly afterwards, the EU and the USA in the TTC (Trans-Atlantic Trade and Technology Council) agreed to team up more closely in international standardization including in the ITU. We see the emergence of a strategic partnership approach. However, despite the push to move Chinese vendors out of EU telecoms infrastructure, by 2022 these still have a very substantial presence according to (Strand Consult, 2022).

The USA

In the USA telecoms restructuring of AT&T in 1984, dominant open market thinking, the appeal of opening-up to the emerging market of China and vice versa keeping the USA market open, combined with little government interest, led to the significant presence of Chinese equipment suppliers in the American market. Moreover, R&D telecoms expertise such as in Bell Labs and Lucent became acquired by foreign companies (in this case, European ones such as Alcatel, and later Siemens, and then Nokia). Telecom operators in the USA were a mix of USA-originated companies, T-Mobile from Germany, and on a smaller scale, Chinese operators. All in all, strategic autonomy was not on the agenda, neither from the cybersecurity perspective, nor from the economic security perspective.

With the rise of cybercrime, the general vulnerability of American telecoms to state-sponsored cyber-espionage, ransomware, and other forms of cyber-attack from Russia, Iran, North Korea, and China, and notably with the rise of China as economic and military power, 5G and 4G have become top concerns for national security (in the USA the term national security now includes national competitiveness). The USA has clearly earmarked the further development of 5G and especially 6G as a matter of strategic autonomy, which at best, can be pursued in a strategic partnership and for which some politicians have even advocated for autarky. Further, ideas have been tabled on next-generation telecoms as open source, a take-over of European companies, and more recently in the TTC, teaming up with Europe, and a strong domestic Next G initiative (ATIS, 2022).

The political rhetoric on the 6G shows the strategic intent but a sense of realism is also present. Namely given the erosion of the industrial and research capability and capacity in the USA (an exception being the strong position in advanced semiconductor design for telecommunications) effectively 6G industrial initiatives are assumed to be open to at least a European presence. As is the EU, the USA is starting to pursue a strategic partnership approach. Make no mistake, however, to assume that strategic teaming-up means identical strategic intent. In international relations, there is fundamentally never an identity of sovereignty between countries. When push comes to shove, such as in a semiconductor supply crisis, despite partnership the US government will prioritize US industry. The EU is matching this in a much weaker form only, however, in the much softer formulated EU Chips Act.

China

In China, since the early 2000s, an explicit policy has been pursued to capture (mobile) telecoms as a strategic industry. China has managed to create space for its own 4G development and with significant state subsidies and impressive entrepreneurship and technical skills, companies such as Huawei and ZTE and China Mobile rapidly grew from minnows to world players. By 2015, these players jointly actually managed to capture the largest share of the world market including the European market and a significant part of the USA and Canadian and Indian markets. In the domestic market, they were overly dominant as access for European and other players remained difficult and relatively small. The EU and the USA for a long time underestimated the rise of these Chinese suppliers, firstly, not realizing what this did to erode their national competitiveness, and secondly, not seeing the influence of the Chinese state eroding their national security. In hindsight, China clearly and consistently pursued and continues to pursue—successfully—an autarky approach to strategic autonomy. China is explicit about its intention for dominance in 6G (China State Council, 2015) though the quest for autarky may for years be hampered by foreign dependencies (Rühlig, 2023).

Relatedness of Policies

Strongly related to telecommunications policy are cybersecurity, cloud, data/AI, IoT, and semiconductor policies. In most of these areas, in the EU at least, the approach to strategic autonomy has been based on risk management, which as mentioned above, is rather soft in terms of guarantees for sovereignty. Gradually, however, harder strategic autonomy elements have been appearing in at least some of these policies. For instance, in the ICT supply chain requirements for cybersecurity are being stepped up in the EU Cyber Resilience Act. These requirements are inspired by the EU 5G Security Recommendation and will imply stronger and independent controls on vendors, including those of IoT and other connected devices. Ensuring cybersecurity is thereby less left to third-country control. In addition, high-security cloud requirements are being developed by ENISA, the EU's cybersecurity agency which may include a greater degree of cloud (and thereby data) localization and staff security scrutiny. However, the final form of these requirements was not yet clear at the time of writing.

International Relations and Technology

Before addressing how the interplay of geopolitics and 6G technology may result in four different futures, we briefly digress and discuss how international relations scholars view technology.

In international relations thinking—grossly simplified—the main schools are the realists, liberalists, and contingency-thinkers. Realists consider that the international system of states is basically an anarchy of states. This does not mean that there is chaos but rather that there is no overarching authority. Moreover, states are captured in a 'security dilemma', ever mistrustful of the intentions of foreign states, having to rely on self-help, and likely pre-emptively having to arm themselves. This line also fits in with global or regional hegemonic thinking (Mearsheimer, 1994; Waltz, 2010).

Liberalists consider that there is more to world order than states alone. International organizations and other actors (e.g., private sector, NGOs, or as relevant here, the global tech community) also play a role in international relations. Collaboration between states is possibly and in fact, quite likely out of 'self-interest rightly understood' (de Tocqueville, 1864). Contingency thinking considers that international relations between states 'depend' on history, the identity of states as historically formed and in relation to the 'socialization' between states, as developed over years and in all forms of international relations (cf. the establishment of international institutions/governance post-1945 which was strongly influenced by the traumas of the two World Wars).

The evolution of international relations in relation to technological trends has been occasionally investigated, but generally technology has for a long time and for most international relations scholars been seen as an exogenous factor and mainly as a factor of warfare. The exception is the thinking about the Internet—famous is the Declaration of the Independence of Cyberspace that states *'Governments of the Industrial World [...] You have no sovereignty where we gather'* (Barlow, 1996).

Recently, however, an emerging paradigm, which is still ill-defined, is 'techno-politics'. This has grown out of Science and Technology Studies and takes seriously a two-way interplay of technology and (international) politics (Eriksson & Newlove-Eriksson, 2021). Examples of such techno-politics are where cyber-weapons are used nowadays by maleficent or rogue states to destabilize the established international system of states, creating a 'sovereignty gap' (Kello, 2017); or where a techno-state exercises a degree of surveillance with digital technologies such as mobile phones and smart cameras that could be considered by other countries as an infringement of human rights and even get raised to the level of the UN as a crime against humanity; or where control of electronic ID technology by the private sector straddles into the realm of the sovereign functions of the state (Timmers, 2022a).

Geopolitics Versus Technology, 2035 in Four Scenarios for 6G

It is against the backdrop of sovereignty-threatening forces that we can map the interplay of sovereignty and technology and consider a number of futures or scenarios. While in the following section we approach these futures for digital technologies in general, the reader may keep in mind what this may mean for 6G. The mapping onto 6G futures will then be dealt with in the later sections. Before moving into the detailed scenarios, let's first present a policy narrative on the future of 6G in a geopoliticized world.

A Policy Narrative on Geopolitics Versus 6G in 2035

What will the future of 6G look like when threats to sovereignty and strategic autonomy are on the rise? We can imagine geopolitics in the future to range from an 'anarchy of states', in which sovereign states never can trust each other, and in which they resort to self-help, and may even be at war; to global collaboration that respects the sovereignty of individual states. Future 6G technology can range from fragmented with closed solutions to fully interoperable and open. See Fig. 10.1.

The actors steering 6G's future are states, digital giants, 6G suppliers, 6G user companies, the technology community, the academic world,

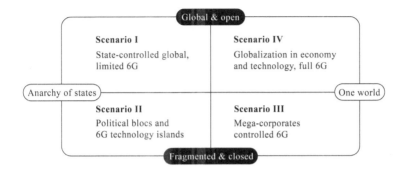

Future of 6G, determined by the fight between: (Geopolitics) & Technology

Fig. 10.1 Geopolitics versus technology

and civil society. Last but not least, there is also a vast army of cyber-criminals. That geopolitics is a major driver for the future of 6G is evident. Many argue that states have a natural tendency to not trust each other. Others are always a threat to survival.

The promise of 6G technology is another major driver. 6G is a huge opportunity and solution for many global problems provided it is integrating, pervasively available, inclusive, and open. Many argue that this is the natural direction for 6G to develop. However, counteracting forces are at work such as global threats of climate change and cybercrime and these can motivate states to collaborate. On the other hand, there is the quest for dominance of large tech companies, who drive 6G toward proprietary oligopolistic configurations. States and large companies may be fighting to control technology, but they can also collude for world dominance.

We can imagine four futures in the fight between geopolitics and technology (see Fig. 10.1, the scenarios are spelled out in detail in the next section):

- *Scenario I*: when states are on a knife's edge, yet they support or force their companies to continue developing global standards and where not possible, they limit 6G.

- *Scenario II*: when global standardization and global collaboration fail, the result is fragmented 6G in a fragmented world.
- *Scenario III*: when states are willing to collaborate, but a few global and dominant companies slice up the 6G space, the result is a 6G of corporate islands.
- *Scenario IV*: when states collaborate and global standardization functions, we could get a global, open 6G in a global, open world.

Which one of the scenarios is most likely to win? Which one do we wish to win? Many will probably say that fragmentation, both geopolitically and technologically, is most likely (scenario II). Many will probably say that a global, open 6G in a global, open world is the most desirable (scenario IV). What if we want to avoid scenario II and maximize the chances of scenario IV? The key is to understand the forces at work and steer them in the right direction by developing our capabilities and capacities, and control 'our' 6G strategic autonomy.

Current Stage?

Let's spell it out. Today, we are probably closest to scenario I. Tensions between states are on the rise and technology is becoming increasingly weaponized. To keep this trend in check, we need to renew cooperation on 6G standards in a balance of countries, companies, the tech community, and civil society. We need new security technology and/or architectures in 6G that safeguard the core of government information and still enable global connectivity and secure end-to-end information flows. Who can do this? The core is a coalition of like-minded countries, companies, and academics (from the USA, the EU, Korea, and Japan) but with openness to include India and, in selected areas where they see a common interest, China.

We also see large corporates increasingly expanding from cloud and platforms into 5G and 6G, involved in a power grab on all the required technologies, from AI and crypto to cables and satellites, from the meta-verse to digital health, retail, and logistics. 6G risks becoming corporatized and fragmented into technology islands. To keep this trend in check, we need the global tech community to team up with innovative emerging disrupters and risk capital, supported by open-minded states to define open standards, make open tech available, and address global challenges with 6G.

Next Steps to Avoid Scenario II and Maximize the Chances for Scenario IV?

What can we do today to best move toward scenario IV, a global open 6G in a global open world? The action must be to *strengthen and expand today's collaborations of the willing that subscribe to the two-track agenda of open global collaboration and open 6G*. What are such collaborations? In the global R&D community, open collaboration used to be supported by the EU Horizon 2020 R&D program, in both telecoms and information technology. This has helped to create a global community of specialists in future technologies and in standards-setting with corresponding forms of global governance. On the telecoms side, the main long-running collaboration platforms for these communities are 3GPP, GSMA, and ITU, while more recently platforms have been emerging around open RAN. These platforms have got and still get their input from large-scale R&D projects (a long tradition pre-6G and continued for 6G, such as in the University of Oulu-led 6G Flagship and many projects co-financed by the EU), private and public–private deployment pilots, and regional but fairly open standardization organizations (ETNO in the EU). On the IT side, the main long-running forms of global governance likewise are still with us from the early Internet times, notably IETF, W3C, IASB for technical architectures and specifications, with the ITU playing a growing role in IT standardization, and ICANN in the management of the Internet domain name system. A plethora of other global alliances exist in the world of AI, cloud, IoT, and even secure computing, which are all potentially relevant for the future of 6G.

The default mindset in this broad community has been and still largely is to enable global open specifications for each generation of 3G, 4G, and 5G, as well as for 6G. Even if much intellectual property is protected by patents for commercial reasons, the assumption of a common global specification for 6G is still driving this 'tech and business community'. One would say then that this broad technical and standards and business community is then the natural anchor for these collaborations of the willing who subscribe to the two-track agenda of open global collaboration and open 6G.

Implications

To assume the above will happen, however, would be naïve: these communities too are pulled into the maelstrom of geopoliticization. The growing concerns about sovereignty and strategic autonomy are also being felt in these forums. The suspicion of the hand of the state behind Chinese companies, which would steer their involvement in global standardization and international R&D collaboration has led to a pushback (Sheehan, 2021; Yan et al., 2019). Various studies document the growing influence of China in standardization, whether in 3GPP or in ITU (Baron & Kanevskaia Whitaker, 2021; Bruer & Brake, 2021). China indeed pursues a very active and comprehensive 5G and 6G leadership policy, as argued above. This is closely linked to its geopolitical ambitions (China State Council, 2015). The EU has become aware of the growing capture of its standardization efforts by foreign companies, and the state actors that are possibly hiding behind them and has come forward with a new approach to standardization that should better safeguard EU strategic autonomy (European Commission, 2022). The USA has been ramping up its involvement in global standardization and expressed willingness to team up with the EU (US & EU, 2022), while launching 6G R&D collaboration, the Next G Alliance, with a strong 'American leadership' ambition (ATIS, 2022). The EU's Joint Undertaking on Smart Networks and Services states to 'foster Europe's technological sovereignty in 6G' (European Commission, 2021). For national regulatory initiatives across the world, see (Serentschy et al., 2022) in this volume. The national and regional 5G/6G R&D and standardization initiatives are increasingly rivalrous. They are being complemented by trade and foreign investment restrictions that cast a dark shadow on the prospect for continued global collaboration. The realistic prospect at the time of writing is of increasing geopolitical tensions that are fueled by the Russian aggression against Ukraine and mainland China's threats against Taiwan and by rising geopolitical rhetoric.

The international academic, technology, standards and business community must go through a self-assessment on how to best handle geopolitics if it is to stay relevant as a platform toward open global 6G. We accept the premise that social construction and technological construction are related, viz. (Cohen, 2019; Lessig, 1999; Timmers, 2022a). Therefore, this should not only be a governance self-assessment but also a technological assessment, answering the question: which 6G technical architecture(s) will enable global collaboration that respects geopolitical concerns about strategic autonomy? In this respect, lessons can be learned from 5G security, that clearly did not live up to this expectation. The issue is that, while 5G was already being rolled out, concerns were raised, initially from the USA, that working with Chinese equipment suppliers such as Huawei would pose a national security risk, due to possible influence by the Chinese state such as on the early stage technology development or on the actual 5G operation. The EU subsequently agreed on a 5G Security Recommendation that would enable to do at least a cybersecurity risk assessment. This experience shows that such security concerns were raised only at a very late stage of 5G evolution, probably due to the lack of earlier involvement of governments. 6G should not fall into the same technical and governance traps that led to this 5G security problem (Timmers, 2020).

A second collaboration that may contribute to open global 6G has been hinted at above. This is where policymakers of 'like-minded' countries (currently the USA, the EU, Japan, South Korea, and possibly India), despite potential differences of views such as on state surveillance (c.f., Voelsen, 2022), discuss aspects of 6G collaboration. Such discussions are happening largely in bilateral dialogues, sometimes with more concrete policy collaboration such as between the EU and the USA in the TTC. An international platform of these 6G policymakers does not exist, though. Provided these policymakers have a clear view on which parts of 6G are highly sensitive to sovereignty concerns, they can issue an open invitation to other countries to join efforts on those aspects of 6G that are not sensitive to sovereignty concerns.

An example would be collaboration on pilots of global interest (e.g., global logistics, public health, climate change) provided 6G is properly technically architected. What does this mean? The 6G academic and technical community needs to urgently develop the characteristics of these parts of 6G. This would likely include compartmentalizing information which is of national security interest. Providing this assessment would be a very important contribution to the techno-politics as a recent and still-evolving perspective on international relations.

Four Scenarios for 6G in Detail

The 1648 Treaty of Westfalen established the system of sovereign states, which consists first of all of the idea that the international order is based upon states and secondly that each state is sovereign and has a sovereign. The latter at the time was a king or emperor but this has evolved via popular democracy into what we now see as the government of a state. So, Westfalen has evolved into the international system of sovereign states, in the West generally internally legitimized by popular democracy, and internationally externally legitimized by treaties, international law, and a plethora of international bodies, most prominently the UN, a rich mix of hard and soft power, and, importantly, having evolved way beyond the idea that the international system of sovereign states is an anarchy of states that in principle are always at war. However, in today's world we do see a harking back to this proto-Westphalia model of sovereign states that principally cannot trust each other and have to rely upon self-help (Waltz, 2010). Here, when we say 'Westfalen' we mean the whole range of geopolitics (i.e., international relations) from Westfalen-1648 to a global world in which sovereignty is compatible with global collaboration.

A Borderless World

Digital technologies largely are constructed as if the world were without borders. From the perspective that social construction and technological construction are related, it is understandable that the until recently dominant paradigms (social constructs) of a global, open, liberal market economy, open research and innovation, academic freedom, and restrained states, have led to such technology. However, and importantly, there is rarely a technical reason that technology must be open, interoperable, uniform, and globally standardized. We can therefore also take technology here on a scale from fragmented to uniformed (on purpose written in this way, that is, formed to be non-fragmented).

This leads to four scenarios as in Fig. 10.1, which, for the purpose of presentation, we label in a somewhat sloganized way as: state-controlled global but limited 6G; fragmentation into political blocs and 6G technology islands; 6G controlled by mega-corporations; globalization in economy and technology, that is, a full, open, and global 6G.

One future may be that geopolitics leads to strong polarization between states ('back to Westfalen'). Alternatively, global cooperation may win in the long run, perhaps under the pressure of the existential global challenges of climate change. In the world of technology, one future is that tech and service companies compete with largely proprietary monolithic solutions. It would be likely then that markets would be served by a few mega-corporations. Another technology future is one of global standards, interoperability, with many building blocks where companies compete through technological and service innovation and possibly disruptive new technologies.

The Role of State Control

Not all technology architectures will fit neatly with strong state control, e.g., non-permissioned distributed ledger solutions in telecoms. Conversely, not all state governance will allow for all technology architecture to develop unchecked. A global world that—hypothetically—would agree on strong fundamental rights including privacy, will wish to keep in check or perhaps even ban privacy-invasive network management architectures. Sovereignty, or more generally, the social constructs that we associate with states, are therefore also conditioned by the way technology is 'constructed', that is, put into technological architectures, specifications, and ultimately coded into hardware and software. Finally, this interplay is neither static nor completely known in advance. Antici-patory regulation is an approach to deal with both the dynamics and the innovation in this interplay (see Serentschy et al., 2022).

In quadrant I, we find a globally open 6G where states are on a knife's edge. Telecom companies have been ordered by their governments to ensure global interconnectivity. Military experts exercise oversight in standardization forums. No information flows can be trusted and there is constant suspicion that implementation is vulnerable to intrusion, back-doors, etc. This necessitates new technical solutions to ensure end-to-end security. Companies will compete globally with extensive backing from their home countries. They will not shy away from using any means to make foreign government buy their gear. The telecoms business may flourish but overheads will be large. 6G strategic autonomy will be a 'militarized' strategic autonomy.

Fragmented 6G

Fragmented 6G in a fragmented world (quadrant II) represents the failure of years of global telecoms and global collaboration in 4G and 5G. The telecom industry has limited markets and seeks value-added in diversification and application, stimulating thriving adjoint indus-tries and entrepreneurship. There is rapid progress in technology within geopolitical blocs provided that these have sufficient scale. Where they do not have this, government subsidies must cover shortfalls. Governments will allow telecom oligopolies to have enough scale. Along the same lines, telecom companies are likely to be eaten by adjoint companies (e.g., cloud providers) to achieve economic scale and protect the state or coalition of states. 6G strategic autonomy means 'fortress strategic autonomy'.

In the third scenario (quadrant III) 6G technology is fragmented, yet there is global collaboration between countries. Mega-corporations have de facto taken over telecoms sovereignty from states and created their own technology islands, likely integrating much more than a traditional telecoms infrastructure (e.g., also satellite systems and cloud platforms and proprietary large-scale AI-as-a-service). Government policies have little effect on these mega-corporations or are absent due to their power and their integration into governmental processes. The '6G strategic autonomy is void'.

A global, open, and full 6G in a globally open world (quadrant IV) is the continuation of the 4G global standards and assumes to counter the risk of security-induced fragmentation of 5G/6G. It requires a careful reflection on 6G architecture to combine openness and security. New security technologies, such as embedded quantum encryption and end-to-end system integrity control with blockchain safeguard the core communications of the state. This requires a new multilateral public–private organization for a balanced and trusted relationship between global 6G companies and governments worldwide, and an understanding that sovereignty battles are better fought elsewhere than in 6G. That is, hard power in other areas is balanced by soft power agreements in 6G, mediated by international diplomacy. Telecom companies will flourish but need to be extremely politically savvy. 6G strategic autonomy delivers a common good, namely 'global 6G peace'.

Discussion and Conclusions

Let's take scenario IV as the desired one, that is, a full and open and global 6G, with global availability and functionality that allows for innovation and inclusiveness. It may be a dream, but only the future will tell. What can be done—in the modest realm of those that are directly involved in 6G—to increase the chances of this scenario becoming a reality? 'Modest', because the likelihood that this scenario will materialize is influenced to a great extent by 'big geopolitics' on which the 6G community of policymakers, industry, and academics may have little influence.

How do we arrive at recommendations on what to do? One way is to look at the other three non-desirable scenarios as destinations to avoid and consider the drivers that push the future in any of those. These are the negative recommendations. Another approach is to consider the drivers of specific positive qualities of the full 6G scenario that create a pull to move in that direction. These are the positive recommendations.

To make this somewhat more concrete, for instance, we would want to avoid fragmentation and securitization (or worse, militarization) of 6G as that will block the road toward realizing a full, global, and open 6G. Openness would be a positive quality of full 6G, so we would want to promote that. This then leads to recommendations as in Table 10.1. The recommendations are labeled with 'to avoid' the bears on the road and with 'to promote' progressing in the desired direction.

We now conclude with several practical and cross-cutting research recommendations toward full, open, and global 6G, for governments, industry, and academia.

For Policymakers

Above all, policymakers would have to promote *open, global, and full 6G as a political vision*, implying that policymakers mobilize their full, and extensive toolset for their 6G industrial policy. This would recognize that digital industrial policy is as much about geopolitics as about the industrial ecosystem and business economics, which is the modern view (Timmers, 2022c).

Governments therefore need to also promote this vision at the global level such as in the UN Global Compact (https://www.unglobalcompact.org/), or minimally to do so multilaterally. A case in point is to build on the Declaration for the Future of the Internet (https://digital-strategy.ec.europa.eu/en/library/declaration-future-internet) and international cooperation such as the EU's Global Gateway (https://ec.europa.eu/info/strategy/priorities-2019-2024/stronger-europe-world/global-gateway_en), such that other countries can join in when they recognize this vision as the better alternative to any other 6G future.

The implication is to develop an *integrated policy* in at least two respects. Firstly, combining traditional 6G policy (R&D and deployment incentives, market regulation) with international/foreign policy actions such as trade policy, strategic international standardization, technology diplomacy, and international or multilateral cooperation. Secondly, to collaborate across user sectors in order to leverage economic, societal, and

Table 10.1 Recommendation toward open, global, and full 6G

Issues	Recommendations			
	Policymakers	Industry	Academia	Civil/tech communities
To avoid Fragmentation	Pro-active policy on open and full 6G and 6G global governance	Open technology development	Early research into fragmentation risks	Common interest and technological solutions
Securitization	Improve engagement on national security interests	Sovereignty-by-design as a business model	Develop sovereignty-by-design approaches	Recognize legitimate concerns
Lock-in	Anticipatory regulation	Value-added and innovation-driven business models	Early monitoring of emergence of lock-in	Early-on discuss, signal and prevent lock-in
Private–public disconnect	Institutional capability and capacity for integrated policy	Re-design public involvement in standardization	Research multi-stakeholderism in geopolitical world	Defend global multi-stakeholder participation
Democratic deficit	Bring 6G in the public debate	6G as theme of industry ESG-D[a]	Define democracy-related harm	Wide public communication
To promote Openness	6G open-source initiative	Free IP	Architect openness vs core of sovereignty	Accommodate widest range of views

(continued)

Table 10.1 (continued)

Issues	Recommendations			
	Policymakers	Industry	Academia	Civil/tech communities
Global standards	Redefine govt. involvement in standards	Redefine govt. involvement in standards	Positive examples of open standards in use of AI and crypto in 6G	Propose and assess global tech standards
Global usage	Multilateral and global fora on common good causes (health, logistics, climate etc.)	Run large-scale pilots in those global use cases	Assess importance and nexus of sovereignty concerns	Mobilize global membership
'Smart' policy	Combine open technology policy with inclusive international policy	Strengthen policy department and share strategic thinking on ESG	Build the case for ESG to include open, human-rights-respecting 6G	Engage from own agenda with industrial and sovereignty agendas
Global commons	Diplomacy and common public interest cases	Common cause use cases	Research techno-politics	Common cause use; responsible technology-social construction

[a]Responsibility for ESG-D = Environmental, Social, Governance, and Democracy

democratic benefits of 6G. A particular important case is 6G military-civil collaboration within public–private relationships (given that open, global, and fully 6G must enable strategic autonomy).

Policymakers should be well-placed to define and initiate *common good global 6G use cases*, such as in global logistics, environmental monitoring, circular economy, and global public health. Collaboration on practical, highly relevant use cases, motivated by 'self-interest rightly understood' (de Tocqueville, 1864) is fertile ground for growing global trust. It is far from evident that policymakers can come forward with integrated, anticipatory, techno-politically and techno-industrial relevant policies. The road to full and open 6G requires breaking down policy silos and policy creativity, and perhaps even a degree of policy experimentation, something that is not in the DNA of traditional policy(law-)makers. *Institutional capability- and capacity-building* is therefore a must and can be stimulated by a close cooperation between policymakers, industry and civil society, while with academia new approaches to flexible regulation— that take account of the interplay between highly dynamic technology development and the more stable governance of 6G—will have to be developed.

Two more specific policy challenges immediately relate to full, global, and open 6G, namely strategic standardization and open-source policy. Governments will have to develop new forms of *engagement in international 6G standardization* in order to safeguard their sovereignty interest, yet not destroy global openness or constrain 6G functionality. Such engagement must reflect the nature of technology and technology architectures that are embedded in 6G or are the foundation of 6G. It will be important to consider legitimate needs to protect the core of government information or ensure the resilience of critical services. Open source is likely to play a central and positive role in this future of 6G. However, governments have traditionally stayed away from the world of open source, for good reason—to not interfere with the powerful innovation and common good which stems from the collaboration of the open-source community, but at times also to bad effect, such as being exposed to cybersecurity risks and corporate capture of open source. Governments will have *to develop and internationally align 6G open-source policy* in order to both avoid such risks, yet ensure that the open-source momentum is maintained notably also from less-powerful contributors.

For Industry

Industry must find new approaches to deal with sovereignty concerns without losing its momentum toward open, global, and full 6G. One approach is to *develop 6G tech alliances that have a political, industrial, and technological anchor* (Timmers, 2022b). That is, these tech alliances should be driven by a political framing (namely, the political vision as mentioned above, as and when it shapes), which provides the strategic intent; industrial involvement as a delivery-/results-oriented platform with a strategic plan that reflects the strategic intent; and a clear technology focus in order to both advance the technology frontier yet remain practical. An example would be AI for 6G linked to *advancing global common good causes* such as sustainable global logistics or preventative global health. Obviously, alignment with governments and civil society is then also possible. A particular responsibility must be taken up by industry to support policymakers, civil society, and academics in ensuring that there is *democratically responsible 6G* in development and usage, or to put it more strongly, that 6G does not lead to democracy-related harm,[3] such as its misuse for citizen surveillance, manipulation of public opinion, or cyber-undermining of society and democracy.

For Academics

This chapter introduced some challenging concepts that must be developed in order to ensure that the movement toward full, open, and global 6G is not stopped in its tracks. This includes *sovereignty-by-design*, which—for legitimate and democratic sovereignty—implies *citizen rights-by-design* (including privacy-by-design) as well as security-by-design. The academic and tech worlds need to develop architectures and individual technologies that respect and reinforce the social constructions of legitimate, democratic sovereignty and of rights.

---

[3] For democracy-related harm, see e.g., Robertson (2022).

There is a long-lasting debate about what citizens' rights are and in international law even human rights do not encompass all that is covered by rights charters such as the fundamental rights that are part of the EU Treaties, let alone political visions such as the EU's Declaration on Digital Rights and Principles (https://digital-strategy.ec.europa.eu/en/policies/digital-principles). Nevertheless, on some rights there is widely shared common ground such as on privacy and rights derived from trustworthy AI.

On a somewhat more reflective level, the academic world must develop thinking on *techno-politics*, i.e., where international relations (IR) embrace technology as an endogenous rather than exogenous force. 6G is par excellence the test case, but more than mere academic interest such conceptual thinking is necessary as an alternative to arms race realist, powerless idealist, or vulnerable contingent IR thinking.

Finally, there is the *trilemma* of (economic) globalization, sovereignty, and democracy, meaning that most but not all three can be realized at the same time, as posited by (Rodrik, 2007; Stein, 2016). If this holds true, one could argue that democracy will suffer from the path toward open, global, and full 6G as of the three it is the one that has the least powerful champions behind it. Whether this is correct and how it should be anticipated and perhaps alleviated, is a challenge for academics to investigate. One way may be to strengthen the global common goods case as suggested by (Stein, 2016), namely for 6G itself to be sustainable and to be a credible contribution to sustainability, as developed in 6G visions, such as (Yrjölä et al., 2020). It could be argued that ignoring the planet-sustainability of 6G would have to therefore undermine sovereignty. Namely, firstly, climate change goes way beyond and is more powerful than any individual country. Lack of 6G sustainability would put the sovereignty of any country at risk. Secondly, sustainability is becoming a profound and cross-cutting requirement (from semiconductors to services to applications). Not mastering 6G sustainability means weakening one's strategic autonomy, i.e., one's capabilities, capacities, and control, and thereby putting sovereignty at risk.

For Global Civil Society and the Tech Community

Finally, global civil society has a particularly important and triple cross-cutting role. Firstly, to defend *causes for the global common good*, societies' specific interest agenda (such as climate, democracy, but also the planet-sustainability of 6G itself, etc.) would be to bring governments, industry, and academia around the table together with civil society and the tech community. Secondly, and related to this, global civil society should provide a *meeting place to grow global collaboration and trust*. Thirdly, to *reduce the risk of democratic deficit* that has been highlighted above, global civil society should become public communicators about the benefits and the risks of full, open, and global 6G.

The global tech community is somewhat less easily identifiable and perhaps more fluid than the other stakeholders. Nevertheless, they were present at the start of the Internet and much of the open software and open hardware developments that will continue to provide building blocks for 6G. The global tech community has a particular responsibility to *bridge technological construction and social construction of 6G*. That is, to develop technological solutions that are compatible with the social constructions in the table above such as promoting openness and avoiding lock-in, or, one of the major challenges, to enable in technological and governance terms sovereignty-by-design.

On Methodology

The analysis of participation is based on https://cordis.europa.eu. A typical Cordis query is:

QUERY = contenttype = 'project' AND frameworkProgramme = 'HORIZON' AND relatedRegion/region/euCode = 'CN' AND ('telecoms' OR '4G' OR '3G' OR '5G' OR '6G' OR 'wireless' OR 'telecommunications' OR 'broadband' OR 'cloud' OR 'mobile' AND 'communications'). Table 10.2 gives the results, where this QUERY is amended or modified as indicated in the first column.

Table 10.2 Participation of Chinese companies in EU-funded projects

EU R&D programme	FP4	FP5	FP6	FP7	Horizon 2020	Horizon Europe	Total
Period	1994–1998	1998–2002	2002–2006	2007–2013	2014–2020	2021–	
China	0	2	20	23	46	2	93
Huawei-China	0	0	3	1	3	0	7
Huawei	0	0	3	9	25	10	47

Note Participation in EU-funded telecom/broadband/cloud

Bibliography

ATIS. (2022). *Next G Alliance*. https://www.nextgalliance.org/.

Barlow, J. P. (1996, February 8). *A declaration of the independence of cyberspace*. Electronic Frontier Foundation. https://www.eff.org/cyberspace-independence.

Baron, J., & Kanevskaia Whitaker, O. (2021). Global competition for leadership positions in standards development organizations. *SSRN Electronic Journal*. https://doi.org/10.2139/SSRN.3818143.

Biersteker, T. (2012). State, sovereignty and territory. In W. Carlsnaes, T. Risse, & B. A. Simmons (Eds.), *Handbook of international relations*. Sage.

Bruer, A., & Brake, D. (2021). *Mapping the international 5G standards landscape and how it impacts U.S. strategy and policy*. ITIF. https://itif.org/publications/2021/11/08/mapping-international-5g-standards-landscape-and-how-it-impacts-us-strategy.

China State Council. (2015). *Made in China 2025*. http://english.www.gov.cn/2016special/madeinchina2025/.

Cohen, J. E. (2019). *Between truth and power: The legal constructions of informational capitalism*. Oxford University Press.

Cowhey, P., & Aronson, J. (2017). *Digital DNA*. Oxford University Press.

de Tocqueville, A. (1864). Comment les AméricainsCombattent L'individualisme par la doctrine de l'intérêt bien entendu. In *De la démocratie en Amérique* (tome II, pp. 198–203). Michel Lévy.

Eriksson, J., & Newlove-Eriksson, L. M. (2021). Theorizing technology and international relations: Prevailing perspectives and new horizons. In *Technology and International Relations* (pp. 3–22). Edward Elgar.

European Commission. (2021). *The smart networks and services joint undertaking. Shaping Europe's digital future.* https://digital-strategy.ec.europa.eu/en/policies/smart-networks-and-services-joint-undertaking.

European Commission. (2022, February 22). *New approach to enable global leadership of EU standards promoting values and a resilient, green and digital Single Market.* https://ec.europa.eu/growth/news/new-approach-enable-global-leadership-eu-standards-promoting-values-and-resilient-green-and-digital-2022-02-02_en.

European Commission, & European External Action Service. (2017). *Resilience, deterrence and defence: Building strong cybersecurity for the EU.* https://eur-lex.europa.eu/legal-content/EN/TXT/?uri=JOIN%3A2017%3A450%3AFIN.

Gallie, W. B. (1956). Essentially contested concepts. *Proceedings of the Aristotelian Society, 56,* 167–198.

Kello, L. (2017). *The virtual weapon and international order.* Yale University Press.

Lessig, L. (1999). *Code: And other laws of cyberspace.* Basic Books.

Mearsheimer, J. J. (1994). The false promise of international institutions. *International Security, 19,* 5–49.

Ministère des Armées | Ministère des Armées. (2022). https://www.defense.gouv.fr/.

Ministry of Defence, UK. (2022). *GOV.UK.* https://www.gov.uk/government/organisations/ministry-of-defence.

Ostrom, E. (2015). *Governing the commons: The evolution of institutions for collective action* (Canto classics ed.). Cambridge University Press.

Robertson, V. H. S. E. (2022). Antitrust, big tech, and democracy: A research agenda. *The Antitrust Bulletin, 67*(2), 259–279. https://doi.org/10.1177/0003603X221082749.

Rodrik, D. (2007, June 27). *The inescapable trilemma of the world economy.* Dani Rodrik's Weblog. https://rodrik.typepad.com/dani_rodriks_weblog/2007/06/the-inescapable.html.

Rühlig, T. (2023). *China's digital power.* DGAP. https://dgap.org/en/research/publications/chinas-digital-power.

Serentschy, G., Timmers, P., & Matinmikko-Blue, M. (2022). Toward anticipatory regulation and beyond. In *The changing world of mobile communications, 5G, 6G and the future of digital services.* Palgrave.

Sheehan, J. F. M. (2021, December 23). *How U.S. businesses view China's growing influence in tech standards.* Carnegie Endowment for International Peace. https://carnegieendowment.org/2021/12/23/how-u.s.-busine sses-view-china-s-growing-influence-in-tech-standards-pub-86084.

Stein, A. A. (2016). The great trilemma: Are globalization, democracy, and sovereignty compatible? *International Theory, 8*(2), 297–340. https://doi. org/10.1017/S1752971916000063.

Strand Consult. (2022, December 14). *The market for 5G RAN in Europe: Share of Chinese and non-Chinese vendors in 31 European countries.* Strand Consult. https://strandconsult.dk/the-market-for-5g-ran-in-europe-share-of-chinese-and-non-chinese-vendors-in-31-european-countries/.

Timmers, P. (2020). There will be no global 6G unless we resolve sovereignty concerns in 5G governance. *Nature Electronics, 3*(1), 10–12. https://doi.org/ 10.1038/s41928-020-0366-3.

Timmers, P. (2022a). The technological construction of sovereignty. In *Perspectives on digital humanism* (pp. 213–218). Springer. https://doi.org/10.1007/ 978-3-030-86144-5_28.

Timmers, P. (2022b). *Strategic autonomy tech alliances.* FEPS Strategic Autonomy Series. https://www.feps-europe.eu/attachments/publications/ 220331%20final_strategic%20autonomy%20tech%20alliances-3a.pdf.

Timmers, P. (2022c). *Digital industrial policy for Europe.* https://cerre.eu/public ations/digital-industrial-policy-for-europe/.

US and EU. (2022, May 16). *U.S.-EU joint statement of the trade and technology council.* U.S. Department of commerce. https://www.commerce.gov/news/ press-releases/2022/05/us-eu-joint-statement-trade-and-technology-council.

Voelsen, D. (2022). *India as an ambivalent partner in global digital policy.* https://www.swp-berlin.org/publikation/india-as-an-ambivalent-partner-in-global-digital-policy.

Waltz, K. N. (2010). *Theory of international politics* (1st ed.). Waveland Press.

Wikipedia. (2022). *Force de dissuasion.* Wikipedia. https://en.wikipedia.org/w/ index.php?title=Force_de_dissuasion&oldid=1126093041#cite_note-4.

Yan, Y., Wong, S.-L., & Liu, N. (2019, July 4). Huawei founder predicts internet of things is next US battle. *Financial Times.* https://www.ft.com/ content/716181ce-9bd8-11e9-9c06-a4640c9feebb.

Yrjölä, S., Ahokangas, P., & Mainmikko-Blue, M. (Eds.). (2020). *White paper on business of 6G* (White paper). University of Oulu.

Part IV

Implications for the Future

11

A View to Beyond 6G

Seppo Yrjölä, Marja Matinmikko-Blue,
and Petri Ahokangas

Alles war Märchen, alles war um eine Dimension reicher,
um eine Bedeutung tiefer, war Spiel und Symbol.
(Herman Hesse, Steppenwolf)

S. Yrjölä (✉)
Centre for Wireless Communications, University of Oulu, Oulu, Finland
e-mail: seppo.yrjola@oulu.fi; seppo.yrjola@nokia.com

Nokia, Oulu, Finland

M. Matinmikko-Blue
Infotech Oulu Focus Institute and Centre for Wireless Communications,
University of Oulu, Oulu, Finland
e-mail: marja.matinmikko@oulu.fi

P. Ahokangas
Martti Ahtisaari Institute, Oulu Business School, University of Oulu, Oulu,
Finland
e-mail: petri.ahokangas@oulu.fi

© The Author(s) 2024   **285**
P. Ahokangas and A. Aagaard (eds.), *The Changing World of Mobile Communications*,
https://doi.org/10.1007/978-3-031-33191-6_11

Assumptions Identified for Mobile Communications

In retrospect, mobile communications can be explained to have developed in an evolutive manner from 1G to the current 5G, as described in preceding chapters. To chart future 5G evolution and 6G, this book applied the technology foresight perspective, which focuses on science, technology, and innovation to make better-informed policy decisions (Pietrobelli & Puppato, 2016), thereby following the technology enablers, regulatory delimitations, and business and other societal and environmental phenomena associated with 6G. To look beyond 6G with a futures research approach, one must therefore pay attention to the weight of the past in mobile communications, the push of the presently recognizable trends, events, and preferences, but especially, emphasize the pull of the future to achieve plausibility (Inayatullah, 2008). For what we label beyond 6G, there will be no single inevitable future, but rather a set of numerous alternative futures. Thus, the future of mobile communications should not be considered evolutive or predictable, but malleable.

To look beyond 6G, this chapter applies a futures research approach that builds on causal layered analysis (Inayatullah, 2019; Inayatullah & Milojevic, 2015), following four steps of inquiry:

Identified assumptions > Presented visions > Transformed futures > Implications for policy

We identify three types of *assumptions* behind the analysis. Ontological assumptions are assumptions regarding the reality faced in the research. Epistemological assumptions are associated with human knowledge, or what forms valid knowledge, whether it can be known, and how a researcher can get it and transfer it. Axiological assumptions concern the level of influence of the researcher's values on the research process, or what is essential and valuable in the research (Burrell & Morgan, 2005). For assumptions, we list the key identified assumptions for the analysis of 6G and beyond. To understand the *visions* related to beyond 6G, we will present the key 6G visions and frameworks by China, Europe, South

Korea, Japan, the USA, and ITU-R. Regarding *transformed futures*, we analyze the presented 6G visions and identify them from legitimation perspective alternative routes to beyond 6G. Finally, we provide policy implications for developing future mobile communications.

Building on the previous chapters, we identify the following assumptions regarding the changes toward future mobile communications generations:

- *From technology-centricity and service-centricity toward human-centricity.* The definition of mobile communications generations has evolved from technology-centric definitions toward service-centricity in 5G and toward human-centricity for 6G and beyond.
- *From technology push toward pull from social and environmental goals.* Up to 5G, the traditional innovation process in mobile communications can be characterized as a technology push from technology and equipment vendors toward operators and end-users. With 6G, new demands for social inclusivity or privacy, security, and safety up to national sovereignty, as well as environmental pressures, have raised triple bottom line sustainability to a driver for developing 6G as a general-purpose technology and an ecosystem-wide effort. This is also expected to continue in beyond 6G.
- *From international to national and local communications toward focal communications.* The provisioning of mobile communications services is changing from international and national operators' mass-produced and top-down offered services toward tailored local communications in 5G and 6G, e.g., with the help of softwarization, virtualization, cloudification, and network slicing. However, it is envisaged that in beyond 6G communications, focally provisioned bottom-up-built personalized on-demand services will emerge.
- *From quality of service and quality of experience toward immersion.* The utilization of mobile communications services has been by provisioning-defined quality of service or utilization-based quality of experience in up to 5G communications. In beyond 6G communications, immersive extended reality, holographic communications, and the metaverse(s) require novel types of quantification for the quality of utilization and experience.

- *From ubiquitous connectivity toward ubiquitous intelligence.* With the convergence of artificial intelligence and other new capabilities like sensing with mobile communications, the assisting and automating role of these capabilities in up to 5G communications is expected to become augmenting in 6G, which means that the nature of communications will change from an availability challenge into what the degree of intelligence or other integrated capabilities available for use is.
- *From human–machine interfaces toward transhumanism.* The traditional device-based use of mobile communications in up to 5G networks is expected to change with new human–machine interfaces like virtual glasses or haptic communications in 6G. For beyond 6G communications, implanted sensors or devices enhance human capabilities and give rise to the emergence of transhumanism, the integration of humans and machines, but also new moral, ethical, and value-related concerns due to the presence of artificial intelligence.

Visions for Future Mobile Communications

This section summarizes the key findings of the recent national developments focusing on the 6G futures envisioned by government initiatives from China, Europe, Japan, South Korea, and the USA. The recent ITU-R activities on IMT toward 2030 and beyond, including 6G, are discussed.

China

In China, mobile communications technologies have become closely tied to national issues of development and prestige, and wider strategic infrastructure and digitalization initiatives such as China standards 2035, belt and road, digital silk road, and made in China 2025. The ministry of science and technology (MOST) has constituted a working group for 6G research, development, and policymaking, consisting of key research institutes and enterprises. In 2019, the ministry of industry and information technology (MIIT) established the IMT-2030 promotion group as the main platform for gathering China's industry and academic forces to promote technological research, advance international dialog and cooperation, and develop a national standardization strategy. In 2020, the Chinese government introduced a subsidies and stimulus package worth RMB 10 T until 2025, focusing on the evolution of 5G, artificial intelligence, data centers, and smart manufacturing. A Chinese 6G white paper on 6G vision and candidate technologies was released in 2021 (China's IMT-2030 (6G) Promotion Group, 2021). The vision, called 'Intelligent connection of everything, digital twin,' envisioned an intelligent era of society built on balanced high-quality social services, scientific and precise social governance, and green energy-saving social development. The vision urges the establishment of new technological industries for high-quality economic growth, driven by the shift from physical products to digital services. The imbalance in wealth and demographics was seen to anticipate changes in social structure. Moreover, a more diversified and flattened governance structure was found to demand scientific and precise governance powered by digital twinning and AI to make timely accurate decisions and respond to real-time topical events. Eight usage scenarios were discussed: the proliferation of intelligence with a ubiquitous smart core; immersive cloud extended reality; digital twinning; holographic communications; converged communication and sensing; sensory fusion; intelligent interactions of feelings and thoughts; global seamless nationwide coverage; and a cross-cutting theme of multilateral network security. China's IMT-2030 (6G) promotion group determined the antecedents of the successful 6G development as follows (China's IMT-2030 (6G) Promotion Group, 2021):

- To ensure the successful commercial deployment of preceding 5G.
- Introduction of native AI intelligence and computing awareness.
- Expansion to higher spectrum bands and bandwidths such as THz and visible light communications.
- To further improve the efficient use of all the spectrum resources via refarming, aggregation, and sharing.
- Expand the coverage ubiquitously on land, at sea, in the sky, and in space.

Europe

In 2020, the European Commission (EC) set extensive political goals to boost economies and competitiveness: green deal; fit for the digital age; an economy that works for people; a stronger Europe in the world; promoting our European way of life; and a new push for European democracy (EU, 2020). It is planned that European geostrategic and global approach initiatives will be pursued via partnerships with like-minded countries and regions. The objectives of the established smart networks and services joint undertaking (SNSJU, 2021) program with the Euro 0.9 B budget are to safeguard industrial leadership and foster technological sovereignty in future 6G. The key means to achieve the goals are the research and innovation program leading to conception and standardization around 2025 and preparatory actions for early market adoption of 6G technologies by the end of the decade. The initial work program has strategic aims to advance an open strategic autonomy via human-centric technologies and innovations. It is envisioned that Europe will become the first digital-led circular, climate-neutral, and sustainable region, leveraging Europe's technological innovation advantages in digital and future emerging technologies (SBS JU, 2021). This is planned to contribute to several key technology policies: green deal; resilient communication privacy and security; AI, data & cloud computing; blockchain technology; high-performance computing; the Internet of things; and microelectronic components. From a services and applications perspective, it is foreseen that the program will impact communication and sensing fusion, immersive environments, digital twinning, and holographic communication. The programs will be measured via democracy, ecosystem, innovation, and sustainability key value indicators:

- Democracy: privacy, fairness, digital inclusion, and trust.
- Ecosystem: sustainability, business value, economic growth, open collaboration, and new value chain.
- Innovation: safety, security, resilience, regulation, responsibility, and energy consumption.
- Seventeen United Nations sustainable development goals (UN SDGs).

The European-level 6G visions (Uusitalo et al., 2021) and initiatives have recently been complemented by several national 6G visions including 6G Flagship in Finland (Latva-aho & Leppänen, 2019), Technology futures in the UK (Ofcom, 2021), and Six questions about 6G whitepaper in Germany (Bayern Innovativ, 2021).

Japan

The beyond 5G promotion consortium (B5GPC, 2020) was established in 2020 by the government, academia, and industry for information sharing and promoting Japanese first-mover advantage in 6G (The Japanese ministry of internal affairs and communications, 2020). The three policy objectives set for the launched program were global first, the creation of an innovation ecosystem, and strengthening competitive advantages in R&D through concentrated allocation of resources (The Japanese ministry of internal affairs and communications, 2020). Japan sets numerical targets for the infrastructure market share (30%) and the share of the number of standard essential patents(10%). The key means of achieving the objectives are to promote global harmonized standardization and policies, collaborative research, and bidirectional globalization.

While targeting the export of its own technologies, Japan also aims to establish a hub of excellence to co-create values for society by calling together researchers and practitioners and their novel visions, as well as technological innovation. In the Japanese 6G vision, it is anticipated that novel mobile communications architecture will contribute to the new sustainable value creation stemming from low energy consumption, improved security and reliability, autonomy, scalability, and advancing 5G performance in data speed, capacity, latency, and density (The Japanese ministry of internal affairs and communications, 2020). From a societal sustainability perspective, it is projected that 6G will develop into a social infrastructure that integrates cyberspace with the real world. In this knowledge-intensive 'Society5.0,' real-time data will be available to all people safely and without an impact on the global environment. In

2021, the National Institute of Information and Communications Technology (NICT) published a beyond 5G/6G white paper, discussing the creation of three scenarios toward 2035: *the cybernetic avatar society; the city on the moon; and transcending space and time* (NICT, 2021).

South Korea

In 2021, the program was launched by the South Korean ministry of science and ICT (MSIT) to accelerate the digitalization of industries, support productivity and economic growth, and 'transform our economy from a fast follower into a pace setter[1]' (MSIT, 2021). MSIT has prioritized 6G, AI, and cybersecurity as strategic technologies in the digital sector. It is planed that the established program will contribute to digital inclusion, digital education, and the digitalization of enterprises. The goal is to gain leadership in international 6G standardization and patenting and be the first in line to launch a 6G trial in 2028, utilizing government-supported public–private partnerships. For the first 5-year period, the government is investing KRW 200 billion, focusing on strategic priorities consisting of space communications, new spectrum (THz) and antennas, ultra-precision, AI, reliability, and the improvement of mobile communications performance KPIs.

South Korea established the 5G Forum Korea as a non-profit organization to promote the evolution and convergence of the extended global ecosystem, especially in the context of Industry4.0. Their 6G vision is based on the selected three drivers of the future society: cleanness and safety; sustainability; and fairness and transparency. The projected key usage scenarios are:

- Internet of inclusive education and experience
- Human augmentation for health
- Sustainable automation in industry and the workforce
- Ubiquitous artificial intelligence in transportation and public safety

---

[1] https://www.korea.net/NewsFocus/policies/view?articleId=187613.

It is envisioned that these usage scenarios will deliver a truly immersive experience, connected real-time intelligence, and interaction between the physical and digital worlds, built on the distributed data-centric network and service infrastructure (5G Forum, 2021).

USA

In the USA, the *Clean Network Initiative* was established in 2020 to address the envisioned cybersecurity threats related to data privacy, security, human rights, and democracy (The United States Government, 2021). To advance future wireless technological leadership, the alliance for telecommunications industry solutions (ATIS) established the NextG alliance (NGA), leveraging sector efforts in 2021 (ATIS NGA, 2021). The NGA program is organized in six working groups: applications; greenG; national goals; societal and economic needs; spectrum; and technology. The developed national roadmap exhibits six objectives for 6G mobile communication networks:

- Resilience, security, privacy-preserving, safe, reliable, and available for private, business, and governmental users.
- Applicability to critical infrastructure, national security, and the military.
- End-to-end cost-effectiveness.
- Supporting life-improving value creation via transformative forms of human-to-human collaboration and human–machine and machine–machine interactions.
- Leveraging artificial intelligence to improve robustness, performance, and efficiency.
- Augmented intelligence with increased flexibility, performance, and resilience built on ultra-reliable low latency communication, multi-sensing, distributed cloud, and virtualization technologies.
- Carbon neutrality by 2040 via 6G energy efficiency and the use of ICT as an enabler.

The NGA's key objective is to advance North American ICT ecosystem leadership over the next decade, encompassing research and development, manufacturing, standardization, and market readiness (ATIS NGA, 2021). In December 2021, the urgency of resilience and sovereignty was stressed by the three cybersecurity senate bills focusing on mobile communications networks, particularly future 6G deployments, and the national cyber literacy campaign (The Senate of the United States, 2021). Societal sustainability and digital inclusion were addressed in August 2021 via the Infrastructure Investment and Jobs Act, which includes USD 42 billion in investments in broadband coverage in unserved and underserved areas.

Global ITU-R

The international level joint process for IMT toward 2030 and beyond, which corresponds to 6G, is underway at the ITU-R working party 5D (ITU-R WP5D). The technology trends have been identified and published in *Future Technology Trends Towards 2030 and Beyond* in late 2022 (ITU-R, 2022). Work on the recommendation about the framework for IMT is ongoing and will be completed in June 2023 with the presentation of new usage scenarios for 6G. Work at the ITU-R WP5D on the framework for the future IMT systems involves member states and organizations worldwide to contribute to forecasting driving factors such as user and application trends, use cases, usage scenarios, and capabilities.

The first report published by the ITU-R in 2022 (ITU-R, 2022) identifies the following services and application trends for IMT toward 2030 and beyond:

- Networks supporting enabling services that help steer communities and countries toward reaching the UN SDGs.
- Increasing customization of user experience with user-centric resource orchestration models.
- Localized demand–supply–consumption models.

- Community-driven networks and public–private partnerships with new models for service provisioning.
- Strong role of networks' vertical and industrial contexts.
- Lowered market entry barriers by the decoupling of technology platforms, allowing multiple entities to contribute to innovation.
- Empowering citizens as knowledge producers, users, and developers, contributing to human-centered innovation.
- Privacy influenced by increased platform data economy or sharing economy.
- Monitoring and steering of the circular economy, including co-creation to promote sustainable interaction with existing resources and processes.
- Development of products and technologies that innovate to zero (e.g., zero-waste and zero-emission technologies).
- Immersive digital realities, facilitating new ways of learning, understanding, and memorizing in different scientific fields.

The spectrum discussions related to 6G are expected to take place at the World Radiocommunication Conference in 2027 (WRC-27) if the WRC in 2023 (WRC-23) develops an agenda item on the IMT spectrum. The actual requirements definition phase will start in the beginning of 2024 and will be finalized in early 2026. The required evaluation criteria and processes will be finalized by the end of 2026. Technology proposals for IMT are expected in 2027–2028, with decisions in 2029.

Transformed Futures for Mobile Communications

This section explores the selected national and regional visions using the causal layered analysis (CLA) method, which articulates alternative perspectives, ideologies, and epistemes, as Table 11.1 illustrates. The futures research CLA method (Inayatullah & Milojevic, 2015) has proven useful in deepening visioning, strategic planning, and policy development. The framework is built on four layers (Inayatullah, 2019). The first is the official unquestioned future documented via the lists of details and contents (*litany* layer). The *social systemic* layer analyzes the litany as consisting of political-environmental, social, technological, economic, and legal (PESTLE) causation and meanings. The third discourse and worldview layer amplifies ideological and discursive assumptions and explores different stakeholder views on the litany and system. Finally, the *myths and metaphors* layer deepens the discovery of the unconscious emotive dimensions. In the CLA process, the four layers of analysis are engaged upward and downward to ensure different ways of knowing and breadth of perspectives. In particular, the ideologies and epistemes of the national and regional 6G vision stakeholders are brought into the worldview and the myth layers. In the final phase, the transformed future perspectives of 6G evolution were developed by reinterpreting the layers, considering the inflection and reconstructing the more visible upper levels of the systems and litany (Inayatullah, 2019) to create reconciled understanding of and visions toward beyond 6G.

Litany Layer

At the surface level, a high similarity of key performance indicators and key value indicators between 6G visions was found, as Table 11.1 summarizes. The *emerging* technologies (Kapoor & Teece, 2021) highlighted in the visions seize the novelty via radical new knowledge and disruptions in terms of improved functionality and the value creation funded. Government funding was widely utilized to cope with initial diversity, high uncertainty, large investments, and the variety of complementary assets to achieve commercialization. The emphasized *enabling* technologies commercializes more horizontally necessitating coordination, a variety of complementary assets and tailored investments in which public–private partnership policies can be utilized to incentivize technological innovations. The envisioned *embedded* technologies were found to further expand the applicability across adjacent businesses. Global harmonization of standards, regulations, and policies will become a key antecedent for interoperability, security, data privacy, AI rights, and novel sharing of economy-based platform business models. In the 6G

Table 11.1 The causal layered analysis of national and regional 6G visions transformed into 6G futures toward beyond 6G

CLA layer	Regional/national 6G visions	Key joint vision elements
Litany of surface-level details of the available 6G visions	• Key performance indicators • Key value indicators • Emerging, enabling, and embedded role of technologies	• 6G general-purpose technology • Global harmonized standardization • Intellectual property licensing policy • Sustainability-driven KPIs and KVIs
Social systemic causation and meanings embedded in the 6G visions	• Capabilities: networks of networks • Leadership • Global vs. national targets • Human- vs. technology-centricity • R&D • Standards • Society • Verticals	• Triple bottom line accounting for social, economic, and environmental sustainability • Trustworthy 6G and stable rules for artificial intelligence and machine learning • Anticipatory regulation promoting open innovation and sustainability
Worldviews and discourse used to legitimate 6G visions	• Competition/ partnerships • National/international • Social perspective and democracy • Growth and innovation	• Ecosystem legitimacy • Empowered human • Citizen-driven • Ethics and morality
Myths and metaphors explaining the 6G visions' deep meaning	• Rights • Level of democracy • Business	• Healing world • Harmonious society advancement

visions, the role of collaboration was emphasized in the early technological innovations. The value creation and capture were seen to increasingly stem from complementary assets and capabilities, shifting from a focal firm-led supply side to dynamic demand and multifaceted business models in platforms and ecosystems. For China, Japan, and South Korea, leadership in standardization is seen as a way to improve quality and increase international competitiveness, while the outcomes are left to the market in the USA.

Social Systemic Layer

In the explored 6G visions, a high similarity of technological capabilities was visible, while differences were found in their exploitation. The competitive free-market approach in the USA emphasizes the role of an enterprise or a focal platform, while the visions originating in Asia underline the role of society and governance. In Europe, global collaboration and the role of environmental, societal, and economic sustainability were central. In general, the nature and speed of technological innovations and the potential disruption related to AI/ML, Web3, and quantum compute were considered to put the current policy and regulatory systems under increasing stress. The common policy concerns of governments about the development of 6G were related to the accelerated global competition and disparity of policies and legislation between the major geopolitical nodes. The development may lead to technological divergence, compartmentalized innovation ecosystems, techno-nationalism, and market protection.

Worldview, Myths, and Metaphor Layers

The underlying assumptions and views in the visions reveal major differences in the legitimation of 6G that can be defined through value creation and value capture (Biloslavo et al., 2020), the selection of the 'right thing to do' (Palazzo & Scherer, 2006), and the socially constructed system of norms, values, beliefs, and definitions within the ecosystem (Suchman, 1995). In the visions originating in Asia, the pragmatic sociopolitical legitimation perspective was visibly adopted, emphasizing innovation compliance with established social rules, regulations, and norms.

In China, passive discursive legitimation was built into the visions, stressing the acceptance and familiarity of existing institutions or the absence of alternatives. The industry legitimation perspective was clearly noticeable in the Japanese and South Korean 6G visions, stemming from the innovation in the industry's institutionalized practices (Kwak & Yoon, 2020). The performative legitimation route in the US visions can be seen to demonstrate the viability of the 6G ecosystem through the processes of strategic action, value realization, adoption, and external intervention. Active legitimation orchestrated by dominant organizations and/or platforms can be built without discursive and performative processes. In Europe, the interaction of the above discussed discursive and performative legitimation was visible in the 6G ecosystem identity construction, founded on an emerging mutual understanding among stakeholders regarding the central, enduring, and distinctive characteristics of the ecosystem value proposition. There was a common understanding that to cope with the liability of newness in 6G and related technological innovation and potential disruption, it would be essential to leverage the interdependencies and resources built alongside collaborative research initiative, regulation, and standardization forums.

Transformed Futures Beyond 6G

Building on the causal layered analysis of the selected national and regional 6G visions, we explored and created new spaces wherein preferable futures visions and strategies could take place (Yrjölä et al., 2022) as Fig. 11.1 summarizes. The analyzed visions shared a characterization of the 6G system as having a ubiquitous economy-wide impact, driving innovation complementarities across industries and application domains and founded on sustained technological improvement across disciplines. Based on the common view on the governing effect on future society, 6G can be defined as a general-purpose technology (GPT) and platform (Teece, 2018). In this vision, the ecosystem legitimacy route will become essential to overcome the liability of newness in 6G.

Collective action among different ecosystem participants through collaborative research projects, trials, and demonstrations to develop a common set of standards that applies to all industries and geographies will ensure consistency, complementarity, extendibility, and economies of scale in the 6G and beyond rollouts. In policymaking, attention

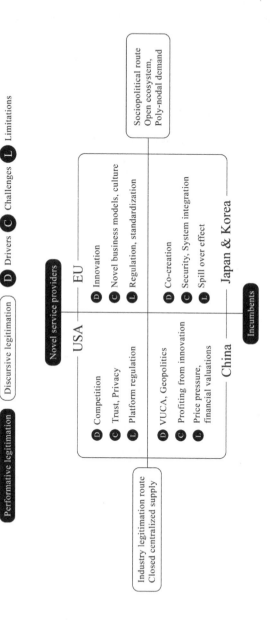

Fig. 11.1 A legitimation view of the worldviews, myths, and metaphors in 6G visions

should be paid to support the value capture for innovators of the critical technological innovations, as the expanded ecosystem will lead to an increasingly complex patent licensing landscape. The sustainability discussion can be seen as moving from an engineering-driven performance discussion to a more holistic triple bottom line visioning, comprising the economic, societal, and environmental perspectives. The introduced complexity to anticipate sustainable development, use of disruptive technologies, and novel innovative business models more flexibly and in a timely manner, with legitimacy and public acceptance, calls for reframed regulation in the form of anticipatory regulation methods. Following the recent discussions and development in harmonized regulation and ethical guidelines for AI, a similar policy for the 6G system can be envisioned. Trustworthy 6G can be characterized as transparent, fair, accountable, robust and safe, human agency and oversight, and private and governed data. 6G futures will extend the role of users from consumer- to producer- and developer-centricity. This emphasizes 6G as an explanatory system that can explain the technical processes and the decisions they make to stakeholders, and the data sets and AI/ML decisions must be documented in a standardized manner to allow for traceability and auditability. The governance mechanisms must support human oversight, and users and developers should be given the knowledge and tools to comprehend and interact with the systems.

Implications for Future Mobile Communications Policy

In the light of the presented discussion, it is envisioned that mobile communications capabilities will play a central role for future societies in different aspects of human and machine life. As futurists, we seek preferred futures but see several bottlenecks and risks facing the development of global 6G and beyond communications that warrant broad policy action and global collaboration.

In recent years, transformative innovation policies have been created to address grand challenges and broader societal goals like those of sustainability or future competitiveness (Diercks et al., 2019; Haddad &

Bergek, 2023). Amidst the current geopolitical tension, we are concerned about the fragmentation of global 6G markets, technologies, and regulations. Transformative innovation policies consider the innovation process and agendas (Diercks et al., 2019) and address system and market failures and developing and expanding new markets, all based on the shared understanding of how change is expected to take place for the preferred future. For policymaking, the identified assumptions, presented visions, and transformed futures give grounds for discussing five intertwined areas of implications for future beyond 6G communications: innovation policy; regulation; sustainability; trustworthy communications; and strategic autonomy and sovereignty.

- *Innovation policy.* Fair or perfect competition never exists. Innovation policies aim to increase companies' and nations' competitiveness both directly and indirectly by affecting firms' intellectual property creation. General-purpose technologies like 6G and beyond require international ecosystemic cross-industry sector innovation. These innovation efforts need to be based on shared goals and expected impacts that enable the creation of a shared vision for 6G and beyond. As future mobile communications technologies are expected to build on the extended use of several complementary technologies such as artificial intelligence, it is of the utmost importance to engage both developers and users of future 6G and beyond in collaboration. We therefore argue for a *global transformative innovation policy* that can push mobile communications ecosystems to deal with system and market failures and address the opportunities for transition and value creation and spillover effects identified for future mobile communications.
- *Regulation.* Although the mobile communications business is highly regulated, the regulative domain for future mobile communications is becoming increasingly complex, especially in Europe, where the mobile communication networks are used to serve specific vertical sectors of society with their own sector-specific regulations. However, the increasing complexity should not create barriers to sustainable value creation or hamper innovation. We therefore argue for *anticipatory regulation*, which defines the rules *ex ante* for developing 6G and beyond, and *ex post* when deploying and using the services. In

practice, the whole regulatory process should be more agile and anticipatory in the context of new technologies, entailing a more proactive,
iterative, and responsive approach to evolving markets' regulation and
emphasizing flexibility, collaboration, and innovation.

- *Sustainability.* Already for 6G, the integrated triple bottom line
  of sustainability (social, economic, and environmental, considered
  in parallel as balanced and uncompromised) has been introduced
  as a new holistic design criterion. What this means in practice
  remains underdefined. 6G and beyond communications can be used
  to solve the grand environmental and social/societal challenges of
  sustainability, provided that a shared vision of 6G exists. On the
  social/societal side, 6G could contribute to fighting climate change,
  decreasing greenhouse gas emissions, or environmental pollution in
  different sectors, but on the social/societal side, the use of artificial
  intelligence may result in new values-based challenges to be solved.
  We argue for *integrated triple bottom line sustainability and resilience*
  when developing future 6G and beyond communications.

- *Trustworthy communications.* The privacy, security, and resilience of
  communications has emerged as important for individual users, organizations, and governments due to the critical role of digital infrastructures and data for modern society and its functions. The embedded or
  inbuilt trustworthiness of communications has thus become a value
  of its own for mobile communications for all its users. Moreover, as
  the amount of intelligence will increase in mobile communications
  and will increasingly be used in mobile communications, the functioning of the systems should be explicable, transparent, accountable,
  fair, safe, oversighted, and controlled by humans. We therefore argue
  for *trustworthy communications as a human right.*

- *Strategic autonomy and sovereignty.* Sovereignty is enabled by strategic
  autonomy in terms of capabilities, capacities, and control regarding
  the economy, society, and democracy. In the context of 6G and
  beyond, sovereignty is enabled and ensured in part by trustworthy
  communications. Digital technologies as a battleground for global
  competition are a source of geopolitical tension and threats against
  societal resilience and diversity; thus, autonomy and sovereignty are

required to maintain competitiveness and sustainability via innovation policies and regulations. We therefore argue for the recognition of
the role of *sovereignty as a basis for fair and legitimate 6G and beyond
mobile communications.*

Bibliography

5G Forum. (2021). *The 5G forum.* http://www.5gforum.org/html/en/mai
n.php.
ATIS NGA. (2021). *NextG alliance (NGA).* https://nextgalliance.org/.
Bayern Innovativ. (2021). *Six questions about 6G* (White paper). https://www.
bayern-innovativ.de/de/seite/whitepaper-six-questions-about-6g.
B5GPC. (2020). *Message to the 2030s.* https://b5g.jp/en/.
Biloslavo, R., Bagnoli, C., Massaro, M., & Cosentino, A. (2020). Business
model transformation toward sustainability: The impact of legitimation.
Management Decision, 58(8), 1643–1662.
Burrell, G., & Morgan, G. (2005). *Sociological paradigms and organizational
analysis.* Ashgale Publishing Company.
China's IMT-2030 (6G) Promotion Group. (2021). *6G vision and candidate technologies* (White paper). http://www.caict.ac.cn/english/news/202
106/P020210608349616163475.pdf.
Diercks, G., Larsen, H., & Steward, F. (2019). Transformative innovation
policy: Addressing variety in an emerging policy paradigm. *Research Policy,
48*(4), 880–894.
EU. (2020). *Strategic plan 2020–2024—Research and innovation.* https://com
mission.europa.eu/publications/strategic-plan-2020-2024-research-and-inn
ovation_en.
Haddad, C. R., & Bergek, A. (2023). Towards an integrated framework for
evaluating transformative innovation policy. *Research Policy, 52*(2), 104676.
Inayatullah, S. (2008). Six pillars: Futures thinking for transforming. *Foresight,
10*(1), 4–21.
Inayatullah, S. (2019). Causal layered analysis a four-level approach to alternative futures relevance and use in foresight. *Futuribles, 12*(3), 123–135.
Inayatullah, S., & Milojevic, I. (2015). *CLA 2.0: Transformative research in
theory and practice.* University of the Sunshine Coast.

ITU-R. (2022). *Future technology trends of terrestrial International Mobile Telecommunications systems towards 2030 and beyond.* https://www.itu.int/pub/R-REP-M.2516.

Kapoor, R., & Teece, D. J. (2021). Three faces of technology's value creation: Emerging, enabling, embedding. *Strategy Science, 6*(1), 1–4.

Kwak, K., & Yoon, H. D. (2020). Unpacking transnational industry legitimacy dynamics, windows of opportunity, and latecomers' catch-up in complex product systems. *Research Policy, 49*(4), 103954.

Latva-aho, M., & Leppänen K. (2019). *Key drivers and research challenges for 6G ubiquitous wireless intelligence* (White paper). http://jultika.oulu.fi/files/isbn9789526223544.pdf.

MSIT. (2021). *The Korean Digital New Deal.* https://digital.go.kr/front/main/eng.do.

NICT. (2021). *Beyond 5G/6G* (White paper). https://beyond5g.nict.go.jp/images/download/NICT_B5G6G_WhitePaperEN_v2_0.pdf.

Ofcom. (2021). *Report: Technology futures—Spotlight on the technologies shaping communications for the future.* https://www.ofcom.org.uk/__data/assets/pdf_file/0011/211115/report-emerging-technologies.pdf.

Palazzo, G., & Scherer, A. G. (2006). Corporate legitimacy as deliberation: A communicative framework. *Journal of Business Ethics, 66*, 71–88.

Pietrobelli, C., & Puppato, F. (2016). Technology foresight and industrial strategy. *Technological Forecasting and Social Change, 110*, 117–125.

SBS JU. (2021). *SNS R&I Work Programme 2021–2022.* https://ec.europa.eu/newsroom/dae/redirection/document/82061.

SNSJU. (2021). *The smart networks and services joint undertaking.* https://digital-strategy.ec.europa.eu/en/policies/smart-networks-and-services-joint-undertaking.

Suchman, M. C. (1995). Managing legitimacy: Strategic and institutional approaches. *Academy of Management Review, 20*(3), 571–610.

Teece, D. J. (2018). Profiting from innovation in the digital economy: Enabling technologies, standards, and licensing models in the wireless world. *Research Policy, 47*(8), 1367–1387.

The Japanese Ministry of Internal Affairs and Communications. (2020). *Beyond 5G promotion strategy: Roadmap to 6G.* https://www.soumu.go.jp/main_sosiki/joho_tsusin/eng/pressrelease/2020/6/30_7.html.

The Senate of the United States. (2021). *H.R.2685-understanding cybersecurity of mobile networks act.* https://www.congress.gov/bill/117th-congress/house-bill/2685/text.

The United States Government. (2021). *The Clean Network*. https://2017-2021.state.gov/the-clean-network/index.html.

Uusitalo, M. A., Rugeland, P., Boldi, M. R., Strinati, E. C., Demestichas, P., Ericson, M., Fettweis, G. P., Filippou, M. C., Gati, A., Hamon, M. H., Hoffmann, M., Latva-Aho, M., Parssinen, A., Richerzhagen, B., Schotten, H., Svensson, T., Wikstrom, G., Wymeersch, H., Ziegler, Z., & Zou, Y. (2021). 6G vision, value, use cases and technologies from European 6G flagship project Hexa-X. *IEEE Access, 9*, 160004–160020.

Yrjölä, S., Ahokangas, P., & Matinmikko-Blue, M. (2022, June). Visions for 6G futures: A causal layered analysis. In *2022 Joint European Conference on Networks and Communications & 6G Summit (EuCNC/6G Summit)* (pp. 535–540). IEEE.

12

Opportunities and Implications Related to Future Mobile Communications

Petri Ahokangas, Annabeth Aagaard, Seppo Yrjölä,
Marja Matinmikko-Blue, Paul Timmers,
Georg Serentschy, Jillian Gordon, Irina Atkova,
Pia Hurmelinna-Laukkanen, Ahmad Arslan, Marika Iivari,
and Oxana Gisca

Life is short,
  Art long,
  Opportunity fleeting,
  Experiment dangerous,
  Judgment difficult.

(Hippocrates)

P. Ahokangas (✉) · I. Atkova · M. Iivari · O. Gisca
Martti Ahtisaari Institute, Oulu Business School, University of Oulu, Oulu,
Finland
e-mail: petri.ahokangas@oulu.fi

I. Atkova
e-mail: irina.atkova@oulu.fi

M. Iivari
e-mail: marika.iivari@oulu.fi

O. Gisca
e-mail: oxana.gisca@oulu.fi

© The Author(s) 2024
P. Ahokangas and A. Aagaard (eds.), *The Changing World of Mobile Communications*,
https://doi.org/10.1007/978-3-031-33191-6_12

Toward Forward-Looking Multidisciplinarity

Digitalization has dramatically changed the role of technologies in communications and information sharing. Over and above this, the ecosystems and value networks that used to be dominant are no longer in

A. Aagaard
Department of Management, Aarhus University, Aarhus, Denmark
e-mail: aaa@mgmt.au.dk

S. Yrjölä
Centre for Wireless Communications, University of Oulu, Oulu, Finland

Nokia, Oulu, Finland

S. Yrjölä
e-mail: seppo.yrjola@oulu.fi; seppo.yrjola@nokia.com

M. Matinmikko-Blue
Infotech Oulu Focus Institute and Centre for Wireless Communications,
University of Oulu, Oulu, Finland
e-mail: marja.matinmikko@oulu.fi

P. Timmers
Oxford Internet Institute, University of Oxford, Oxford, UK
e-mail: paul.timmers@iivii.eu

G. Serentschy
Serentschy Advisory Services GmbH, Vienna, Austria
e-mail: your@advice-serentschy.com

J. Gordon
Adam Smith Business School, University of Glasgow, Glasgow, Scotland, UK
e-mail: jillian.gordon@glasgow.ac.uk

P. Hurmelinna-Laukkanen · A. Arslan
Department of Marketing, Management and International Business, Oulu
Business School, University of Oulu, Oulu, Finland
e-mail: pia.hurmelinna@oulu.fi

A. Arslan
e-mail: ahmad.arslan@oulu.fi

charge of technology evolution. At the center of all this are the customers, users, and developers and their unique needs and specific requirements. Most of the existing digitalization literature explores what is technologically possible or viable, and how future technologies will evolve. This technology push approach typically focuses on new technical developments that the technologists want to see, not what users need. Taking a step further, this book adopted a multidisciplinary and forward-looking approach to make sense of digitalization, especially in the context of future mobile communications, seeing technology as an enabler, regulation as a conditioning and limiting factor, and business as something that will emerge between technology and regulation. From the technological perspective, the chapters presented an outlook for the capabilities needed for various envisioned services enabled by future 5G and 6G. From the regulatory perspective, the chapters discussed the increasingly complex regulatory domain emerging for future mobile communications and presented some solutions for dealing with the complexities of anticipatory regulation. From the business perspective, the chapters provided an outlook on value creation and capture with novel services and business models, discussing the platform-based and ecosystemic features of future 5G and 6G. Finally, the chapters discussed the social and societal context for future 5G and 6G as they concern individuals, businesses, markets, societies, and geopolitics. What remains to be discussed are the implications and opportunities arising from the presented analysis and discussion.

This chapter will discuss the implications for research, management, and policymaking stemming from the preceding chapters. The idea is not to wrap up the chapters; rather, the aim is to identify and map *opportunities* and *implications* for further action for researchers, managers making decisions in the mobile communications context, and policymakers and the authorities.

Opportunities and Implications for Research

This section will identify opportunities and implications from business, regulation, and technology perspectives for researchers interested in future mobile communications-related topics, regardless of their discipline or domain. The following research methodology and research theme-related opportunities and implications are presented for the future:

- *Apply cross-disciplinarity and future research methods.* Understanding and exploring mobile communications as a research domain requires knowledge from several disciplines, ranging from engineering, business, economy, law, international relations, sociology, and psychology to mention only some. The current and emerging phenomena around 5G and 6G call for a combination of scientific knowledge and approaches from several fields that warrant a cross-disciplinary approach and a forward-looking methodology, specifically with qualitative research but also with various multi-method combinations. This also requires the involvement of various stakeholders to open the door for non-dominant voices to be heard when designing the future.

  – What does cross-disciplinarity mean in the future mobile communications context?
  – What theories and approaches are suitable for explaining the phenomena around future 5G and 6G?
  – What are the new cross-disciplinary design criteria for future mobile communications systems and services?

- *Emphasize the triple bottom line of sustainability and sustainable development goals.* Sustainability has emerged as a new design criterion for future 5G and specifically 6G. Green radios with lower energy consumption have long been a research topic, along with how to minimize electromagnetic fields in mobile communications. Yet the triple bottom line of sustainability, covering environmental, economic, and social perspectives, and especially what it means, how it can be reached, and what its measurable impacts are, remain interesting research questions. In addition, different notions of social and societal sustainability in different cultures and government systems provide impetus and importance for research into what social and societal sustainability means, and how it can be achieved. Questions concerning how ethnicity, inclusivity, and human-centricity can be embedded in the development of mobile communications arise.

- What are the new sustainability-driven performance and value-related requirements for 6G and beyond?
- How can 6G and beyond technologies and services contribute to solving sustainability challenges?
- How can ethical principles be introduced to future mobile communications?

- *Examine general-purpose technologies and innovation.* Technologies define the opportunities available for business and society. 6G and artificial intelligence are becoming tethered together, giving rise to a new type of general-purpose technology domain that will have far-reaching impacts on everyday life. However, the outcomes and impacts and the mechanisms behind them, specifically related to this combination of technologies, on individuals, organizations, businesses, markets, and societies, are poorly understood and present opportunities for future research.

- How will 5G, 6G, and artificial intelligence change societies and human life?
- What mechanisms can be used to explain and understand innovation behavior and the benefit for firms and societies?

- *Explore platformization and 'ecosystemization' of businesses.* On the consumer side, platforms have become the mainstream of global business. However, platformization and its related emergence of ecosystems—with traditional business incumbents becoming platform-based—are entering every business field, making multiplatform ecosystems the dominant type of business, represented by owners and complementors. However, theories that capture and explain strategies within such a new platform or ecosystem economy are still in their infancy.

- What kind of strategies emerge in multiplatform ecosystems?
- Will vertical-specific multiplatform ecosystems differ from those of consumer markets?

- What theories can help develop our understanding of strategies employed in multiplatform ecosystems?

- *Explain novel ecosystemic business models for value creation and capture.* The business model has become the contemporary tool for describing and designing businesses in the digital era, with value creation and capture at the heart of the discussion. Related to platforms, ecosystems, and general-purpose technologies, novel business models, and business model conceptions are expected to arise for public and private organizations, but also due to the increasing human-centricity of digitalization for individuals.

  - Will there be specific business models for new services such as network slicing, artificial intelligence, or metaverses in the future?
  - How can we use business model theory to explain firms' strategies and behavior?
  - To what extent can business model theory be used to foresee future strategies?

- *Explore human experience and services. In the future*, empowering experiential citizens as knowledge producers, developers and users will contribute to a process of human-centered democratizing innovation stemming from pluralism and diversity. With the expected emergence of new human–machine interfaces, holographic communications, haptic communications, and metaverses, human experience and empowerment are expected to expand to new heights. Data-intensive, human experience-enhancing services, especially if ubiquitously available, are technically and regulatorily challenging.

  - How should human experience be researched in the context of the metaverse or human–machine interaction?
  - How will human–machine interaction and metaverses change human needs, servitization, *markets, and the consumption of services*?

- *Ensure privacy, security, resilience, strategic autonomy, and sovereignty.* Future mobile communications must be understood and designed to ensure privacy, national and personal security, the availability and integrity of communications and data, and respect for the fundamental rights of the individual and for countries' self-determination.

These principles and procedures must meet the legitimate expectations set by the stakeholders in future mobile communications, but they may also produce differing implications in different contexts, as they may be of concern for nations regarding strategic autonomy, democracy, and sovereignty.

- How can 6G technology and governance be designed and societal acceptance be supported so that they evolve fruitfully together?
- How can we collaborate internationally to reap the full benefits of a global 6G and to avoid a backlash by concerns about sovereignty or fundamental rights?

- *Explore regulatory antecedents and outcomes to develop anticipatory regulation.* The increasingly complex regulatory environment surrounding mobile communications-enabled businesses is becoming challenging to comply with, and in different markets and countries, the regulatory domains may have contradictory requirements, leading sometimes to unwanted or unexpected outcomes. Anticipatory regulation has been seen as an opportunity to develop future regulation, but new knowledge is needed for how to understand and approach it.

- What is the minimum that needs to be regulated?
- How can we examine and anticipate the outcomes and impacts of regulation?
- How can we avoid the negative outcomes and impacts of regulation?

- *Benefiting from innovation and transformative innovation policies.* Like the regulatory domain, the policy domain needs a holistic understanding of how firms and societies can benefit from innovation. As innovation policies should holistically consider innovation activities, their lifecycle, and commercialization efforts, transformative innovation policies should also have predefined and measurable outcomes and impacts. However, theories that explain the benefits of innovation and transformative innovations are yet to be combined.

- What kind of innovations policies should there be for future mobile communications?
- The systemic and complex converging 6G platform and ecosystem provide an exciting research context to study how to profit from

innovation, particularly in relation to the open architecture and open source adopted in 6G.

– How do globalization or global strategies and innovations policies interact within mobile communications?

• Risks, threats, and negative impacts of platformization and digitalization. 5G and 6G are typically associated with progress and development. However, they can be seen as enablers of harmful and even criminal developments in society.

– What could the downsides be of progressing digitalization and advancing mobile communications, and how can we deal with or solve them?

Managerial Opportunities and Implications

The 6G era is characterized by a completely new kind of convergence and complementarities as a multiplatform ecosystem with a central role of general-purpose technologies, which generates new kinds of needs to understand how value is created, delivered, shared, and captured. For business managers, the discussion in the book chapters gives rise to the following observations regarding business opportunities and implications:

• Value appropriation (capture) outcomes may not only be about the immediate profits from innovation, but the benefits may be quite varied, from private to social returns, and they may accrue over time. This gives rise to the following implications for management:

– Innovators may monetize their innovations by paying attention to the appropriability regime covering legal instruments, especially intellectual property rights (IPRs), tacit, hard-to-imitate technologies, and complementary assets like algorithms, data, interfaces, and ready-made components.
– Attention should be paid to the instruments, processes, and outcomes of value appropriation in different areas/markets of future 6G.

- For incumbent firms, the changing role of innovation and ecosystems will entail the need to develop new ecosystem-embracing strategies and business models in their existing markets. The question is about choices:

  - Identify emerging new ecosystems and business verticals in the intersection of different technological domains.
  - Develop hybrid business models that extend from connectivity toward product-service models building on higher 4C layers like context and commerce.
  - Considering the organizational perspective, organize activities outside the focal enterprise to implement new business models, and design and implement the business process across the ecosystem.
  - Bring human-centricity to business model innovation.

- The decoupling of technology platforms will lower the market entry barrier, allowing multiple entities to contribute to the innovations envisaged for 6G. For startups and small and medium-sized enterprises, the exploration of growth opportunities with complementary services will become essential.

  - Adopt complementor strategies for quick market entry.
  - Develop new business models for slicing, secure zones, sensing, and sustainability.
  - Privacy regulation will be strongly linked to the rising trends of the platform data economy, sharing economy, intelligent assistants, connected living in smart cities, transhumanism, and digital twins' reality.
  - Attention should be paid to the value appropriation outcomes of business activities at the ecosystem level.

- In the emerging new ecosystems, understanding management and the organizational models needed for specific business models becomes central for business model implementation and competitiveness. New kinds of governance and organizing logics are needed for the coopetitive context, where collaboration and competition take place on both sides of platforms. This raises several implications:

- Identify the real needs in various vertical and industrial contexts to develop a strong role for 6G.
- Identify the decision-making logic in the ecosystem.
- Identify the fundamental role/position choices that need to be made regarding how to do business in the new ecosystems.
- Avoid siloed thinking in multiplatform ecosystem contexts.
- Prepare for two-sided coopetition.
- Build alternative governance structures for multiplatform ecosystems.

• It is not enough for platform owners and complementors that technical, service, and business infrastructures will exist and be available in the future 6G era. It is essential to consider whether users have real access to these services—that they have the required devices, and that they also know how to use them, as well as the other adjacent/complementary services. The role of developers has been emphasized in emerging 6G (general-purpose) technology contexts. Moreover, openness can benefit original innovators later even if they release (or lose) the technological innovations to the surrounding environment where others exploit it. A deeper understanding of technology in the form of design and development skills such as programming or digital fabrication may also further enhance users' opportunities to play an active role in the ecosystem and make and shape technologies for their personal needs. This also assists users in evaluating and reflecting on the technologies and their role in the user's own life, as well as more widely in society.

- Identify and explore the spillovers and social returns that could become a new source of private returns for the initial innovator.
- Examine non-users and understand the reasons for their exclusion: Is it by their own choice or for another reason?
- Identify who benefits from technology or service use, and how.
- Who experiences value?

- What is the real price of services, and is it worth paying?
- Organize and balance value-related processes to achieve diverse triple bottom line sustainability goals.
- Integrate social and environmental goals into business models.
- Decentralized platform cooperatives will become counterforces to winner-takes-all platform monopolies.

- There will be traditional stakeholders in the future mobile communications ecosystem like mobile network operators, mobile virtual network operators, mobile communications technology providers, and mobile communications equipment providers, as well as different new ones like resource and asset providers, matching and bridging service providers and a variety of new types of service users.

  - Benchmark best newcomer strategies to enter the mobile communications domain internationally.
  - Build customized user experiences with user-centric resource orchestration.
  - Identify new sources of value creation in metaverse variants.
  - New societal models for future service provisioning will emerge, building on community-driven networks and public–private partnerships.
  - Learn the regulations that apply in mobile communications ecosystems.

- It may be expected that with new generations of mobile communications, the commercialization and standardization lifecycle will get more complicated due to the need to integrate/converge new technologies like security-enhancing ones or artificial intelligence to 5G and 6G. This raises at least the following for consideration:

  - Make a clear strategy for contributing to the definition, standardization, and deployment/use of a technology generation.
  - Develop corporate responsibility around future 6G and share best practices.

- Local-demand–supply-consumption models will become prominent in an already globalized world, with opportunities in localized spatial circular economies.

• From the environmental perspective, 6G is seen as a provider of services to help steer communities and countries toward reaching the UN SDGs. However, the UN SDGs should have been achieved by the time 6G enters the market. 6G will offer opportunities for monitoring and steering the circular economy and understanding the big picture of the sustainable data economy. It is also expected that companies will shift the focus, developing products and technologies that innovate to zero, including zero-waste and zero-emission technologies bringing social innovation to the fore.

Implications for Policymaking

The traditional questions for policymakers with a new mobile communications generation at hand have been about how to deal with new spectrum bands. The challenge of finding more spectrum for mobile communications is a fundamental problem for which the solution space differs by the stakeholder. The only way out of this challenge is sharing-based spectrum access, which allows different radio systems to share the spectrum under predefined rules and conditions. The technology for this is already mature; now, it is time to introduce these principles to mobile communications, which has traditionally relied on exclusive long-term spectrum licenses over wide areas.

However, policymaking is not only about spectrum management. How to address the emergence of a large number of local 6G networks deployed by different stakeholders for different types of use faces not only the spectrum challenge but also competition, openness, regulatory, and technology, and innovation policy challenges. These themes extend the discussion to the global scale, requiring harmonization in several fields. Among the most important are: spectrum-related policies and principles; standardization across different technology areas and industries; international trade and fair competition; data ownership, governance, and platforms; privacy, security, and user/consumer rights; innovation policies and international collaboration; and artificial intelligence.

Moreover, digital technologies have become the basis for global competitiveness and geopolitical rivalry. They are becoming the source of geopolitical tension and are being weaponized. One consequence is the fragmentation of regulations. Apart from the increasingly complex regulations, there is now also a geopolitical motivation to turn away from global interoperability and open interconnection. The 'splinternet' is already a fact today. Industrial and innovation policies are becoming more compartmentalized and are driven by geopolitics as much as by national competitiveness and company interests. Trade impediments, data colonialism, and increasing concerns about strategic autonomy and sovereignty all risk contributing to the globalization backlash, creating a real possibility that we will never see a truly global 6G.

However, there are also forces that require global 6G and beyond mobile communications. The most pronounced of these are global grand challenges, notably climate change and the need to achieve environmental sustainability at a global level, global health scares (pandemics), and rampant worldwide cybercrime. Environmental sustainability has emerged as a new design principle for future mobile communications. However, we argue that social and societal sustainability, inclusivity, human-centricity, and human rights should also be considered, as indicated and measured by the United Nation's seventeen social development goals, the UN SDGs. In current discussions, the concept of trustworthy and resilient networks or communications has been related to the privacy, security, safety, and resilience of communications, but the concept could also be seen in a wider context. This gives rise to the following implications for policymaking:

- Despite geopolitical tension, promote open, global, and full 6G as a political vision.
- Collaborate as policymaking communities with researchers, academia, and business to advance common global good use of 6G, addressing global challenges in particular.
- Avoid regulatory and standardization fragmentation by enabling global collaboration on legal and technical matters.

- Develop integrated policy for 6G to achieve economic, social, and democratic progress, respecting human-centricity and human rights and addressing security concerns by design.
- Collaborate internationally to anticipate the future of technologies and their economic and social appropriation, notably on

  - The combination of electronic communications with cybersecurity, quantum computing, cyber-physical systems, and artificial intelligence, as well as on
  - The evolution of platformization and its effects in terms of competition, global inequality, and equity.

Finally, we emphasize that sustainability will be a game changer for beyond 6G. The ways and criteria for how countries are evaluated as pioneers in mobile communications will change drastically. The total amount of consumed mobile data is far from sustainability thinking. Telecommunications policymakers face the challenge of defining new rules and requirements for something that is outside their traditional competence area.

- Develop new metrics and methods to assess environmental, social, and economic sustainability of 6G and beyond (footprint), as well as the use of mobile communications in other sectors to make a positive impact (handprint) on the environment (see, for example, the recent discussions on nature positive strategies by the World Economic Forum) and societies at large.
- Provide visibility for end users about the sustainability impact of their decisions on ICTs.
- Establish the research community's role as the provider of unbiased research results that are the foundations for future policymaking.

Index

A

Advantage x, 33, 49, 74, 80, 115, 141, 142, 144, 150, 153, 193, 194, 210, 212, 230, 233, 290, 291
AI act 129, 195, 205, 206
Algorithm 21, 59, 100, 124, 125, 146, 150, 155, 156
Alliances 22, 76, 174, 256, 257, 265, 276
Anticipatory regulation 9, 237, 240, 241, 246, 270, 301, 302, 309, 313
Application programming interface (API) 71, 73, 124
Appropriability 9, 31, 169, 170, 172–176, 178, 314
Artificial Intelligence (AI) ix, xxiii, 4, 9, 19, 29, 31, 54, 59, 63, 65, 66, 69, 73, 75, 81, 82, 84, 85,
87, 103, 128, 129, 138, 154, 156, 158, 161, 168, 173, 177, 178, 188, 205, 206, 231, 260, 264, 265, 270, 276, 277, 288–290, 292, 293, 296, 298, 301–303, 311, 317, 319, 320
As-a-service 31, 35, 57, 76, 146, 160, 175, 212, 270

B

Benefiting from innovation 167, 169–171, 313
Bit-pipe 148
Broadband 15, 27, 47, 68, 75, 99, 116, 139, 148, 257, 278, 294
Business model x, xiii–xv, xx–xxii, xxiv, 8, 9, 27, 29–33, 35–37, 57, 58, 71, 72, 75, 76, 87, 89, 115, 126, 138–161, 173–175,

178, 187, 191, 193, 194, 199,
202, 207, 210, 211, 231, 240,
296, 297, 301, 309, 312, 315,
317
Business model innovation 58, 152,
153, 157, 194, 315

C

CAPEX 76
Causal layered analysis (CLA) 10,
286, 295–297, 299
Cloud x, 18, 26, 29–31, 35, 36, 49,
52, 55, 57, 60, 64, 71, 73, 75,
124, 144, 146, 147, 150,
155–158, 160, 169, 172–174,
176, 200, 201, 212, 215, 229,
239, 245, 246, 257, 260, 264,
265, 270, 278, 289, 290, 293
Code 176
Commerce 27, 30, 83, 138, 140,
145, 148, 150, 158, 160, 315
Commercialization 14, 126, 152,
176, 296, 313, 317
Community 7, 19, 36, 47, 56,
68–70, 72, 86–88, 96, 99,
101, 104, 126, 127, 129, 139,
152, 156, 175, 261, 262,
264–268, 271, 275, 278, 294,
295, 317–320
Competition xiii, xiv, xxiii, 10, 22,
27, 35, 76, 80–83, 86, 123,
142, 160, 169, 172–174, 177,
187, 191, 192, 198, 199, 204,
208, 210, 222, 224–227, 231,
232, 234–236, 242, 243, 246,
247, 298, 302, 303, 316, 318,
320

Connectivity v, vii, x, xiv, 6, 8, 15,
16, 21, 22, 27, 29, 30, 33, 35,
49–52, 54, 60, 63, 71, 72,
74–76, 99, 117, 122, 127,
130, 138, 141, 143, 145–150,
154–158, 160, 168, 169, 174,
186, 187, 192, 204, 212, 229,
238–240, 242, 264, 288, 315
Content 4, 6, 7, 15, 20, 26, 27, 30,
33, 50, 51, 54, 57, 64, 116,
122, 129, 138–140, 142, 143,
145, 147–150, 155, 157, 158,
160, 187, 197, 199, 200, 208,
212, 230, 235, 239, 296
Context x, xxi, xxii, xxiv, 4, 7, 9, 26,
27, 30–32, 54, 75, 77, 87, 89,
94, 95, 97, 100, 101, 103,
104, 114, 115, 117, 118, 124,
138–140, 144, 145, 148–150,
152, 155, 156, 158, 160, 169,
170, 172–178, 186, 189, 190,
192, 195, 211, 230, 231, 242,
246–248, 292, 295, 303, 309,
313, 315, 316, 319
Core network 15, 16, 21, 146, 242,
256
Cybersecurity 56, 293

D

Data vi, xxii, 4, 15, 16, 18–21, 26,
27, 30–32, 36, 47, 49–54, 57,
59, 60, 63, 64, 66, 67, 69, 72,
73, 75, 76, 78, 81, 82, 85–88,
98, 99, 101, 116, 117, 124,
125, 129, 138, 145–150,
154–157, 169, 172, 173, 186,
187, 190, 192, 196, 198, 199,
201, 202, 204, 207, 210, 212,

215, 229, 233–237, 239, 260,
    289–291, 293, 295, 296, 301,
    303, 312, 314, 315, 318–320
Data Act (DA) 195, 201, 244
Data Governance Act 195, 202
Democracy 56, 79, 80, 87, 103,
    129, 254, 255, 268, 276–278,
    290, 293, 303, 313
Deployment v, vi, 14–16, 25, 27,
    31, 47, 49, 68, 72, 74, 76, 89,
    100, 102, 116, 120, 129, 144,
    150, 152, 178, 189–191, 195,
    207, 212, 215, 222, 257, 265,
    272, 289, 294, 317
Digitalization x, xiv, xxii, 4, 6, 14,
    19, 46, 52, 57, 59, 79, 117,
    138, 142, 152, 154, 191, 198,
    203, 215, 222, 233, 237, 288,
    292, 308, 309, 312, 314
Digital Market Act 188
Digital Service Act 200

E

Ecosystem xiv, xv, xix–xxii, 5, 7–9,
    16, 19–22, 26, 27, 29, 30, 33,
    36, 37, 49, 57, 60, 66, 69, 71,
    72, 74–76, 80, 86, 87, 99,
    100, 114–120, 123–126, 129,
    130, 139, 141, 144–147, 149,
    152, 155–161, 169, 174–178,
    186–189, 192–194, 210, 211,
    215, 222, 225, 237, 239, 242,
    246, 272, 287, 290–292, 294,
    297–299, 301, 302, 308,
    311–317

Edge cloud vii, 26, 49, 52, 76, 156,
    212, 239
Enhanced mobile broadband
    (eMBB) 15, 16, 116, 120, 139
e-privacy 196, 203
EU 2030 Digital Compass 194, 198
European Electronic
    Communications Code
    (EECC) 24, 188, 195, 204,
    205, 228
European Telecommunications
    Standards Institute (ETSI) 20,
    22
Experience xxii, 4, 7, 21, 26, 31, 49,
    56, 59, 61, 63, 65, 67, 68, 70,
    86, 87, 116, 123, 143, 192,
    231, 237, 245, 247, 267, 287,
    292–294, 312

F

Fair, reasonable and
    non-discriminatory (FRAND)
    20–22
Fifth generation (5G) v–vii, ix, x,
    xiii, xiv, xx, xxii, 4–9, 15, 16,
    18–20, 24–31, 33, 35, 46, 47,
    54, 55, 75, 85, 87, 89, 98,
    116, 120, 121, 124–126, 130,
    138, 139, 143–150, 153, 155,
    157, 158, 168, 169, 171, 172,
    174, 178, 186–192, 195–197,
    203, 204, 206–213, 215, 221,
    222, 224, 233, 234, 242,
    257–260, 264–267, 270, 271,
    278, 286–289, 291–293,
    309–311, 314, 317

First generation (1G) 4, 7, 8, 15, 19, 286
Fourth generation (4G) 4, 8, 15, 20, 24–29, 31, 33, 35, 98, 142, 168, 169, 171–173, 221, 224, 257, 259, 265, 270, 271, 278
Future vi, vii, x, xiv, xix, xx, xxiv, 4–10, 18, 25, 32, 46, 47, 49, 51, 54, 56–58, 60, 64, 78, 79, 82, 86, 87, 96, 97, 102–105, 116, 117, 126–130, 139, 141, 143, 145, 148, 150, 152–157, 160, 161, 169, 172, 173, 177, 178, 188, 189, 192, 196–199, 203, 207, 209, 222, 228, 229, 231, 234, 237, 239, 240, 242, 243, 249, 255, 256, 261–263, 265, 269, 271, 272, 275, 286–288, 290–297, 299, 301–303, 309–314, 316, 317, 319, 320

G

General-purpose technology (GPT) 9, 22, 36, 71, 126, 152, 168, 169, 172, 174–177, 287, 299, 311
Geopolitical 5, 8, 9, 19, 57, 74, 75, 78, 80, 86, 89, 129, 241, 256, 266, 267, 270, 298, 302, 303, 319
Global commons 256, 276–278

H

Human–machine interfaces (HMI) 31, 32, 51, 87, 157, 158, 288

Human-centric vi, 6, 116, 126, 130, 158, 160, 198, 210, 290

I

IMT-2020 15, 19, 25
IMT-2030 288, 289
Inclusivity 103, 287, 311, 319
Incumbent 15, 30, 35, 71, 72, 74, 76, 154, 158, 178, 225, 311, 315
Industrial policy x, 9, 211, 230, 242, 246–249, 272
Information and communications technology (ICT) xiv, xxii, 5–7, 49, 53, 96, 99, 101, 102, 113, 114, 117, 118, 153, 194, 211, 224–227, 232, 233, 237, 238, 241–247, 260, 293, 294
Innovation vi, x, xi, xiv, xix, xxii–xxiv, 9, 14, 16, 19, 20, 22, 23, 30, 31, 33, 35–37, 57, 68, 69, 71–76, 80, 81, 83, 86–89, 95–97, 100, 105, 117, 125, 130, 152, 156, 167–170, 172–179, 187–189, 191, 192, 198, 202, 206, 207, 209, 210, 215, 222, 223, 225, 226, 230, 231, 236, 237, 239, 241–249, 269–271, 275, 286, 287, 290, 291, 295–299, 301–303, 311–316, 318
Innovation policy 177, 231, 241, 243, 246, 249, 302, 318
Interface 6, 14, 19, 21, 22, 30–33, 37, 54, 63, 64, 66, 73, 74, 78, 102, 114, 118, 124, 125, 129, 150, 154, 156, 157, 174, 312, 314

Internet of Things (IoT) ix, xxi, 15,
16, 53, 63, 65, 75, 116, 146,
148, 154, 201, 222, 239, 260,
265, 290
ITU-R 14, 15, 17, 18, 24, 25, 96,
98, 116, 139, 153, 287, 288,
294

K

Key performance indicator (KPIs)
98, 127, 292, 296
Key value indicator (KVIs) 127,
290, 296

L

Law 53, 83, 173, 188, 199, 200,
202, 209, 210, 223–226, 228,
231, 232, 234, 241, 247, 268,
277, 310
Legitimacy xxi, 9, 187–189,
192–194, 196, 206–211, 254,
299, 301
Licensing 20–22, 191, 235, 301
Local network 27, 86, 121, 123,
124, 130, 139, 144, 148, 149,
154, 186, 188–192, 207–212
Lock-in 70, 115, 118, 278
Long term evolution (LTE) x, 16,
19, 221, 224

M

Machine Learning (ML) ix, 16, 31,
36, 49, 54, 298, 301
Massive machine-type
communications (mMTC) 16,
116, 120, 139

MNO 24, 27, 30, 50, 76, 123, 124,
139, 148, 154, 155, 189–191,
207
Mobile communications xiv, xxii,
4–10, 14, 15, 19, 20, 22,
24–27, 29, 30, 32, 33, 35,
46–49, 71, 89, 96, 98,
100–102, 104, 113, 116, 117,
121, 124, 125, 128–130, 138,
139, 142, 143, 145, 146, 148,
150, 151, 153, 160, 161,
167–169, 175–177, 186–191,
193, 207, 211, 221, 222, 226,
286–288, 291–294, 301–304,
309–313, 317–320
Mobile network operator (MNO)
xiv, 15, 19, 20, 25–27, 30,
74–76, 101, 102, 121–124,
138, 139, 143–149, 154–158,
160, 172, 186, 190–192, 207,
212, 317
Mobile operator 14, 71, 130, 139,
143, 145, 147, 148, 153, 154,
158, 160, 172
Mobile technology 20
Mobile virtual network operator
(MVNO) 26, 123, 147, 317
Multidisciplinary xxii, xxiv, 4, 5,
187, 241, 309
Multilateralism 254
Multi-level perspective (MLP) 96,
97, 100, 103, 104

N

Network v, vi, ix, x, xiv, xxiv, 4, 5,
9, 15, 16, 18–21, 24, 26, 27,
29–32, 35–37, 47–57, 59, 60,
63–66, 68–76, 78–80, 85,

87–89, 99–101, 118,
120–124, 126, 128–130,
138–142, 144–150, 153–155,
157, 158, 160, 169, 172–176,
186–192, 195–197, 199, 200,
203–207, 209–213, 215, 223,
230, 233–235, 239, 242, 269,
287–289, 293–295, 302, 308,
312, 317–319
Network effect 30, 118, 126
New Radio (NR) 16

O

Open source 35, 52, 71, 73, 88,
259, 275
Operating expenses (OPEX) 76, 153
Opportunity vii, 27, 51, 63, 71, 72,
74, 94, 115, 123, 125, 129,
141, 142, 148, 150, 157, 176,
193, 194, 246, 263, 313
Over-the-top (OTT) xiii, 15, 26,
27, 49, 54, 63, 71, 74, 75, 86,
142, 146, 155, 158, 160, 190,
205, 238

P

Patent x, xxiv, 20, 71, 233, 242,
265, 291, 301
Platform ix, x, xxiii, xxiv, 7–9, 16,
18–22, 27, 29, 30, 32, 33,
35–38, 49, 52, 53, 57, 58, 63,
67, 71–76, 78, 82, 86–88,
116–120, 122, 124, 125, 130,
138, 140, 141, 145–147, 149,
150, 152, 154–158, 160, 161,

169, 170, 173–175, 186, 192,
196, 199, 200, 202, 212, 227,
230, 238, 242–245, 247, 248,
254, 264, 265, 267, 270, 276,
289, 295–299, 309, 311, 312,
315, 316, 319
Policymaking 6, 7, 10, 234, 244,
246, 288, 299, 302, 309,
318–320
Privacy xxiii, 4, 18, 20, 33, 50, 53,
67, 74, 76, 84, 88, 89, 103,
128, 129, 153, 158, 173, 186,
187, 192, 207, 210, 269, 277,
287, 290, 293, 295, 296, 303,
312, 319
Proprietary 19, 22, 65, 124, 173,
175, 263, 269, 270

R

Radio access network (RAN) 16, 21,
265
Regulation x, xiv, xxii, xxiii, 4, 5, 9,
14, 15, 22, 24, 25, 49, 51, 53,
58, 64, 67, 68, 75, 76, 81, 83,
87, 88, 101, 104, 122, 123,
142, 143, 145, 152–154, 160,
173, 178, 187–192, 196, 197,
199, 201, 203, 205, 207,
209–212, 215, 222–227
Regulatory governance 222, 237,
245
Resilience 49, 66, 74, 78, 80, 84,
94, 103, 104, 129, 153, 170,
175, 202, 203, 275, 290, 293,
294, 303, 312, 319

Risk management 200, 256, 257,
260

S

Safety 16, 18, 31, 47, 52, 59, 66,
69, 103, 128, 129, 186, 198,
287, 290, 292, 319
Scenario xi, 6, 8, 10, 14, 25, 47, 52,
60–63, 65, 67, 68, 70–72,
74–80, 83, 86–88, 99, 116,
127, 128, 139, 243, 244, 247,
248, 262–265, 268–271, 289,
292–294
Second generation (2G) 4, 15, 27,
28, 146, 148, 221
Security vi, 4, 18, 20, 21, 26, 33,
52, 53, 72, 74, 76, 81, 84, 85,
89, 103, 128, 129, 153, 157,
158, 173, 186, 187, 192, 198,
203, 204, 207, 210, 222, 234,
235, 239, 254, 255, 257–260,
264, 267, 268, 270, 271, 287,
289–291, 293, 296, 303, 312,
317, 319, 320
Service v, vi, ix, x, xiii, xiv, 4–6, 8,
15, 16, 18, 19, 21, 22, 24–27,
29–31, 33, 35–37, 47, 49–55,
57–61, 63–67, 69, 71–76, 82,
86–88, 99–103, 113–118,
120–130, 138–140, 142,
144–149, 154–158, 160, 169,
170, 173–176, 178, 186–192,
196–205, 212, 215, 223,
229–231, 234–236, 238, 239,
242, 246, 269, 275, 277, 287,

289, 290, 293–295, 302,
309–312, 315–318
Service-level agreement 76
Servitization 52, 66, 130, 312
Sixth generation (6G) v–vii, ix, x,
xiii, xiv, xx–xxii, 4–10, 18,
20–23, 25, 26, 30–37, 46–49,
51–54, 58–60, 63–68, 70,
72–74, 76, 77, 79, 82, 86–89,
96–105, 116, 117, 126–130,
139, 143, 144, 152–160,
168–179, 186–190, 192,
195–197, 206–213, 215, 225,
233, 234, 237, 239, 240, 242,
246, 249, 255–257, 259–273,
275–278, 286–304, 309–311,
313–320
6G and beyond 6, 286, 287, 299,
301–303, 319, 320
Slicing ix, 16, 31, 36, 74, 124, 147,
149, 150, 154, 287, 312
Smart city xx, 47, 67, 69
Sovereignty vii, 9, 10, 19, 103, 126,
129, 153, 211, 228, 229, 240,
241, 244, 245, 247, 253–257,
259, 260, 262, 266–271,
275–278, 287, 290, 294, 302,
303, 312, 313, 319
Spectrum vii, x, xxii, xxiii, 22, 24,
25, 31, 47–49, 54, 71, 75,
138, 139, 142, 157, 172, 173,
191, 205, 207, 210, 289, 293,
295, 318
Spillover effect 121, 123, 157, 177,
302
Standard essential patent (SEP) 20,
22, 291

Standardization 9, 14, 17, 19–21,
    25, 37, 71, 101, 124, 125,
    129, 152, 173, 242, 243, 247,
    254, 257, 258, 264–266, 270,
    272, 275, 289–292, 294, 297,
    299, 317–319
Strategic autonomy 129, 153, 229,
    241, 244, 245, 247, 254–260,
    262, 264, 266, 267, 270, 271,
    275, 277, 290, 302, 303, 312,
    313, 319
Strategic partnerships 255, 256, 258,
    259
Strategy vii, xxiii, 35, 140, 141, 147,
    150, 151, 160, 197, 202, 203,
    210, 233, 289
Sustainability vi, vii, ix, xxii, 53, 56,
    58, 59, 63, 66, 67, 74–84, 86,
    94–104, 115, 126, 128, 129,
    142, 148, 150, 152–154, 170,
    173, 175, 193, 194, 198, 210,
    212, 277, 287, 290–292, 294,
    298, 301–304, 310, 319, 320
Sustainability transition 8, 95–100,
    102–105
Sustainable Development 58, 87,
    94, 95, 98, 153, 301

Technology generation 8, 14, 19, 24,
    27, 146, 152, 157, 317
Third generation (3G) 15, 19, 24,
    25, 27, 28, 221, 257, 265, 278
Transhumanism 53, 58, 63, 78, 88,
    128, 288, 315
Trends xiii, xxi, 8, 25, 47, 48, 53,
    54, 57, 60, 74, 76, 86, 88,

    154, 215, 227, 235, 261, 264,
    286, 294, 315
Trustworthiness vii, 53, 129, 153,
    210, 303
Trustworthy x, 66, 103, 117, 126,
    128, 129, 153, 196, 202, 277,
    301–303, 319

U

Ubiquitous vi, 49, 52, 57, 127, 130,
    138, 144, 145, 153, 221, 222,
    239, 288, 289, 292, 299
Ultra-reliable low-latency
    communications (URLLC) 15,
    16, 116, 120, 139
UMTS 19, 221
User Equipment (UE) 16, 21, 31

V

Value capture x, 7, 20, 35, 71, 76,
    118, 119, 130, 155, 156, 167,
    168, 173, 194, 298, 301
Value chain 8, 27, 28, 35, 57, 70,
    74, 75, 119, 154, 156, 190,
    290
Value co-capture 37, 119, 176
Value co-creation 30, 37, 38, 57,
    114, 116–121, 126, 130, 149
Value creation 7, 8, 32, 33, 56, 65,
    70–72, 74, 75, 113, 115–117,
    119, 120, 123–130, 140, 149,
    150, 152, 155, 160, 193, 194,
    210, 212, 291, 293, 296–298,
    302, 309, 312, 317
Value proposition 30, 36, 115, 120,
    126, 140, 145, 158, 160, 192,
    193, 299

Vertical vi, 10, 15, 20–22, 26, 27, 33–35, 61, 67, 69, 70, 74, 75, 77, 80, 81, 87, 127, 139, 146, 148, 149, 155, 158, 160, 178, 186, 191, 192, 207, 212, 246, 295, 302

Vision vi, vii, 4, 5, 10, 15, 18, 25, 96, 177, 198, 209, 210, 230, 243, 245, 272, 276, 277, 286, 287, 289, 291, 292, 295–303, 319

Volatility, uncertainty, complexity, and ambiguity (VUCA) 32, 46, 150

W

Webscale 71, 76, 84, 156